CROSSING THE HEARTLAND

TOURING NORTH AMERICA

SERIES EDITOR

Anthony R. de Souza, *National Geographic Society*

MANAGING EDITOR

Winfield Swanson, *National Geographic Society*

CROSSING THE HEARTLAND

Chicago to Denver

BY

JOHN C. HUDSON

RUTGERS UNIVERSITY PRESS • NEW BRUNSWICK, NEW JERSEY

This book is published in cooperation with the 27th International Geographical Congress, which is the sole sponsor of *Touring North America*. The book has been brought to publication with the generous assistance of a grant from the National Science Foundation/Education and Human Resources, Washington, D.C.

Manufactured in the United States of America

Rutgers University Press
109 Church Street
New Brunswick, New Jersey 08901

The paper used in this book meets the minimum requirements of American National Standard for Information Sciences—Permanence of Paper for Printed Library Materials, ANSI Z39.48-1984.

Library of Congress Cataloging-in-Publication Data

Hudson, John C.
Crossing the heartland: Chicago to Denver / by John C. Hudson—1st ed.
p. cm. —(Touring North America)
Includes bibliographical references and index.
ISBN 0-8135-1880-6 (cloth: alk. paper)—ISBN 0-8135-1881-4 (paper: alk. paper)
1. Middle West—Tours. 2. West (U.S.)—Tours. 3. Middle West—Geography.
4. West (U.S.)—Geography. I. Title. II. Series.
F355.H74 1992
917.8—dc20
92-10535
CIP

First Edition

Frontispiece: The Platte River east of Kearney, Nebraska. Photograph by James L. Amos for the National Geographic Society.

Series design by John Romer

Typeset by Peter Strupp/Princeton Editorial Associates

Contents

PART THREE

RESOURCES

Foreword

Touring North America is a series of field guides by leading professional authorities under the auspices of the 1992 International Geographical Congress. These meetings of the International Geographical Union (IGU) have convened every four years for over a century. Field guides of the IGU have become established as significant scholarly contributions to the literature of field analysis. Their significance is that they relate field facts to conceptual frameworks.

Unlike the last Congress in the United States in 1952, which had only four field seminars, the 1992 IGC entails 13 field guides ranging from the low latitudes of the Caribbean to the polar regions of Canada, and from the prehistoric relics of pre-Columbian Mexico to the contemporary megalopolitan eastern United States. This series also continues the tradition of a transcontinental traverse from the nation's capital to the California coast.

This traverse from Chicago to Denver crosses the heartland of American agriculture and the rich and scenic variety of the Great Plains. It provides an understanding of the physical and cultural underpinnings of this remarkable region.

John C. Hudson, professor of geography and anthropology at Northwestern University, is the nation's recognized scholar on the cultural and historical development of the American plains.

Anthony R. de Souza
BETHESDA, MARYLAND

Acknowledgments

I acknowledge the dedicated work of the following cartographic interns at the National Geographic Society, who were responsible for producing the maps that appear in this book: Nikolas H. Huffman, cartographic designer for the 27th International Geographical Congress; Patrick Gaul, GIS specialist at COMSIS in Sacramento, California; Scott Oglesby, who drew and oversaw production of the relief art work; Lynda Barker; Michael B. Shirreffs; and Alisa Pengue Solomon. Cartographic assistance was provided by the staff at the National Geographic Society, especially the Map Library and Book Collection, the Cartographic Division, Computer Applications, and Typographic Services. Special thanks go to Susie Friedman of Computer Applications for procuring the hardware needed to complete this project on schedule.

I thank Lynda Sterling, publicity manager and assistant to Anthony R. de Souza, the series editor; Richard Walker, editorial assistant at the 27th IGC; Natalie Jacobus, who proofread this volume; and Tod Sukontarak, who indexed and photoresearched the volume. They were major players behind the scenes. Many thanks, also, to all those at Rutgers University Press who had a hand in the making of this book, especially Kenneth Arnold, Karen Reeds, Marilyn Campbell, and Barbara Kopel. Marlie Wasserman and Kate Harrie of the Press lent a hand (and eye) in proofreading this volume on their native region.

Errors of fact, omission, or interpretation are entirely my responsibility, and any opinions or interpretations are not necessarily those of the 27th International Geographical Congress, which is the sponsor of this field guide and the *Touring North America* series.

Financial support for the preparation of this guidebook was provided by the William and Marion Haas Fund, Northwestern University.

PART ONE

Introduction to the Region

Introduction

Crossing the Heartland traverses six states of the Middle West and Great Plains. Beginning at St. Joseph, Michigan, the itinerary includes Gary, Indiana; Chicago and Rock Island, Illinois; Iowa City, Des Moines, and Sioux City, Iowa; Norfolk and Ogallala, Nebraska; and ends just short of the Rocky Mountain foothills at Denver, Colorado. Except for the major metropolitan centers of Chicago and Denver, however, the focus of the excursion is on the rural landscape rather than the city. Regional terms, such as "Heartland," "Corn Belt," or "High Plains" suggest the distinctive features of this broad stretch of the United States. The physical landscape itself, especially the natural setting as it has been modified by centuries of human use, will be emphasized in the commentary.

Geologically speaking, both the Middle West and the Great Plains are part of the Central Lowland of North America. This largest geologic subdivision of the continent is bordered by the Appalachian Mountains on the east and by the Rocky Mountains on the west. Its distinctive features are its low surface relief and its comparatively low elevation in the middle of a large land mass. The regional label, "Middle West," is a term of human geography that refers to the twelve states in the triangle between Ohio, North Dakota, and Kansas. "Great Plains," in contrast, is a physical-geographic unit, the eastward-sloping surface of relatively higher elevation that borders the Rocky Mountains to a distance of 300 to 400 miles on the east. Our journey will reach the Great Plains in eastern Nebraska roughly where the transition to the Sand Hills begins.

UNDERSTANDING THE LANDSCAPE

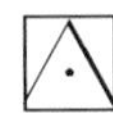

To appreciate the landscape variety of the interior of the United States requires attention to subtle differences. The land is not totally flat in most places, yet no ranges of mountains help the traveler mark passage across the region. Glaciation during the past 100,000 years muted the region's topographic features. Ice sheets a mile or more thick planed off the hilly summits and filled the valleys with debris. Hundreds of millennia in the past the Middle West looked more like Kentucky's or Tennessee's rougher plateaus do today.

Between the forested Great Lakes and the pine-dotted Rocky Mountain foothills almost any route selected for travel will be fringed by patches of woodland. Yet, for most of the distance across the breadth of this land, grasses are considered to be the natural vegetation. Transitions between hills and plains or forests and grasslands are gradual. The zones of transition themselves are often as large as the relatively isolated sections of pure landscape types. Climate and topographic texture define the various combinations of land types in the middle United States, although neither controlling factor exerts so dominant an influence that the other can be excluded if one wishes to understand why the landscape looks like it does.

Broad-leaf deciduous forests dominate the east, grasses blanket the west, and a mixture of coniferous and broad-leaf species covers the north. Those three types of plant cover may be considered in combination with, broadly speaking, two kinds of topography: the Middle West and the Great Plains are either gently sloping or, occasionally, hilly. Grasses typically are found on flat to gently rolling topography where the fires thought necessary for their maintenance spread rapidly. Broad-leaf deciduous forests, dominated by oak, elm, maple, beech, or basswood, occur on all types of topography around the Great Lakes. The species composition changes toward the west in favor of cottonwoods and the more drought-resistant oaks, and in the Great Plains broad-leaf forests are confined almost entirely to flat lands along the major rivers. Coniferous forests are found on the cold, northern edge of the Middle West, where white pine, black spruce, and hemlock thrive; and also on the dry slopes of buttes and mountain foothills in the west, home of the ponderosa pine, cedar, and juniper.

Other differences in biota appear at a more local scale. Grasses grow tall on the comparatively wet Grand Prairie of Indiana and Illinois, but they decrease in height westward from there as the relative balance between precipitation and evaporation of soil moisture changes from a water surplus to a water deficit. Where soil texture is coarse, however, as in the Sand Hills of Nebraska, tall grasses, such as big bluestem, reappear. Conifers and even desert plants take root naturally to dominate isolated, south-facing rocky precipices along the Mississippi River in Wisconsin and Illinois, sites whose microclimates more closely resemble the desert southwest. Ecosystem boundaries are thus shifted east or west, north or south, based on variations in secondary factors such as soil texture or topographic orientation.

Such local exceptions to regional patterns are common in the mid-continent. In fact, variety at the microscale can be more dramatic than the differences one observes at the regional scale. Travelers should take the time to look closely, for no continuously unfolding pageant of landscape change is to be found merely through casual observation. Yet, one can find much of interest here. Those who would just as soon skip this region, fly over it, or cross it entirely at night need only an open mind and a willingness to let their natural curiosities be aroused.

VARIETIES OF ENVIRONMENTS

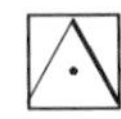

Four major environmental types occur widely in the stretch of country between the Great Lakes and the Rocky Mountains: boreal forests, hilly woodlands, flat to rolling grasslands, and flat woodlands.

Boreal forests are found only in occasional patches in southern Michigan and Wisconsin and are even preserved as rare curiosities in isolated bogs of northern Illinois and Indiana. Minimum winter temperature, rather than cool summer temperature, defines the southern limit of widespread boreal forest. Northern Michigan, Wisconsin, and Minnesota are in the boreal forest zone, meaning that their winter temperatures are often cold enough to kill temperate-zone broadleaf deciduous trees. The boreal zone also has a shorter growing season. Its southern limit marks the northern edge of the agricultural Middle West.

Hilly woodlands are of two distinct types. In the Ohio, Mississippi, and lower Missouri river valleys, rolling hills covered with a dense forest of broad-leaf trees and a thick undergrowth are present-day reminders of how unattractive hilly, broken slopes were to those who sought to make a living from the land. Such lands may have been cleared at some time in the past, but they were used only sporadically and have reverted to forest. South of the limit of glaciation, in southern Indiana, Illinois, and Missouri, the land surface is comparatively rough. Forests there are expanding in acreage today, as more and more marginal farmland goes out of production.

The other type of hilly woodland is found in the western Great Plains, but there the climate is dry and the trees are dry-zone conifers such as the ponderosa pine. One can travel along a river such as the Niobrara in northern Nebraska and remain in a narrow corridor of pine forest, surrounded everywhere by grasses on the uplands stretching back from the steep valley sides. The Pine Ridge in Nebraska, the Black Hills in South Dakota (so named because of the dark hue of the trees contrasted with the adjacent lowland grasses), and numerous isolated buttes from the Dakotas south to Texas are forested, their steep slopes being the primary factor determining the vegetation as woodland rather than grass.

Grasslands, east or west, are either flat or rolling. The terms "prairie" and "plains" are often used interchangeably, although the French "prairie" denotes a vegetation type while "plains" is a descriptor of topography. The confusion is understandable inasmuch as grasses and smooth land are generally found together.

The North American prairie takes on the shape of a giant wedge that covers most of the interior of the continent. The broad side of the wedge adjoins the Rocky Mountains, from Alberta to Mexico. East from there, however, the wedge narrows almost constantly, and reaches an apex along the Illinois and Indiana border. This easternmost projection, known as the "Prairie Peninsula" in Illinois, carries the grassland biome well into the humid, forested east.

The association between smooth topography and grasses is one key to prairie origin. Fires burn rapidly across a smooth surface, but they generally halt at the edge of steeply sloping valley sides. Fire is a necessary part of any explanation of grassland origin in this region, although the relative influence of natural forces (lightning) versus human agency (firing the lands to drive game and increase the nutritional value of grasses) remains a subject for debate.

The other important variable is climate. Prairies cover a region of the Middle West that, although it is supplied with precipitation sufficient to support broad-leaf forests, is especially prone to summer drought. Summertime dominance by the mild, dry breezes of the Pacific air stream in this zone of westerly circulation can produce dessication of plants, thus establishing a precondition for the autumnal prairie fires of the sort that eighteenth- and nineteenth-century Euro-American explorers recorded in their journals.

Regardless of who or what caused the prairie to spread east as far as the southern end of Lake Michigan, the grassy lands are associated with smooth topography and demarcated as a region that has more frequent summer drought than the lands that border it.

Flat woodlands, the fourth environmental type, are primarily associated with riverine locations. The margins of small lakes away from the principal rivers may be considered in the same category. These are comparatively wet environments, subject to seasonal inundations or periodic fluctuations in the water table. Lands in such locations, too wet to permit trees to root, typically have a vegetation of reeds and marsh grass, a condition especially prevalent across the glaciated Middle West where thousands of small, wet depressions dot the surface.

A journey across the Middle West and the Great Plains involves a continuous succession of transitions between these environmental types. Omitting the boreal forest category, which is found only in the north, a transit from Chicago to Denver involves frequent passage from flat, prairie uplands down through wooded bottomlands, and back to the grassy uplands once more, here and there passing tracts of hilly woodland. The trip is marked not so much by a major transition in types as it is by a succession of small transitions, controlled by the topography, soils, and local availability of water.

THE LANDSCAPE'S APPEARANCE

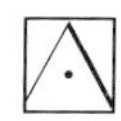

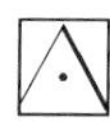

Three underlying factors control broader-scale differences in the landscape's appearance. They are the continental-interior climate, a geological underpinning of horizontal sedimentary rocks, and a nearly constant regional slope.

THE CLIMATE

At a regional scale, as when one observes differences between places, say, 100 or more miles apart, climate is the more important factor influencing vegetation. The species composition changes inexorably as one progresses from the humid east toward the subhumid to semiarid west: grasses change from tall to medium to short; except along the rivers, massive broad-leaf trees are replaced by more stunted forms; tree-forms give way to shrub-forms; water-demanding broad-leaf varieties in the east are replaced by drought-tolerant conifers in the west. In some places soils mitigate the effects of climate; in others topography creates marked varieties in the plant cover within a single climatic zone, but a trip from east to west, down the precipitation gradient, is marked by a decline in the biomass present on the landscape. Species of flora and fauna that have evolved an adaptation to dryland living replace their humid-zone relatives toward the west.

Precipitation decreases from east to west because of continentality: the farther west toward the Rocky Mountains one travels, the more distant is the Gulf of Mexico. Warm, moist air masses move northward from the near-permanent cells of high atmospheric pressure that occur over the subtropical portions of the Earth. The Mississippi River Valley is well supplied with moisture all year from the northward movement of these subtropical air masses, but precipitation drops off toward the west once past the longitude of the western Gulf of Mexico. It is comparatively rare for Gulf air to reach eastern Colorado, slightly more common for it to arrive in central Nebraska, and so on, with increasing frequency toward the east. Thus, the gradient of annual precipitation is arrayed in an almost perfect east–west alignment across the Great Plains and western Middle West. Travel across Nebraska or Kansas and you will find that annual precipitation decreases about one inch for every 20 or 30 miles of travel toward the west (approximately one centimeter every 15 kilometers).

The Gulf of Mexico is not the only source of moist air, of course. The Rocky Mountain region gets its rain and snow from the Pacific Ocean's air masses, which, when they are lifted up the west-facing slopes of the mountains, produce abundant precipitation. But the Great Plains and the Middle West lack topographic barriers that force air currents aloft and thereby precipitate rain from the clouds. The lift necessary to cool the air and make it rain is provided, instead, by dense, cold air layers that move south from Canada. Mountains of air instead of rock, these cold air masses resist the tendency to mingle with the warm, and thus they retain their defining characteristics. Warm air is forced up above cold just as it is up the side of a mountain. Middlewesterners are used to talking about weather fronts—zones of contact between cold, dry air from the north and warm, moist air from the Gulf along which the warm air is lifted, cooled, and thereby drained of its moisture.

Variable weather is associated with these fronts, which draw the unlike air masses into a cyclonic, inward spiral of circulation. A warm and cloudy day is typically followed by rain and then often cooler and drier conditions. As the joke is told in North Dakota or Kansas or Illinois: if you don't like the weather, just wait a few hours and it will change. Periods of weather variability are usual, rather than exceptional, so the well-equipped traveler should carry clothing for the variety of conditions that prevail even within a single season.

Being virtually landlocked, the Middle West and the Great Plains are subjected to the influence of air masses from the north, west, and south. In the spring, especially, when cold, dry air is still coming south, and warm, moist air has begun to penetrate northward, tornado conditions can result. Violent storms, dry spells, and heat waves are some of the consequences of being in the intersection zone of these disparate outside influences. Also, with no nearby ocean to cool it in summer or help warm it in winter, the mid-continent's annual cycle of temperature is one of broad swings.

GEOLOGICAL UNDERPINNINGS

The flatness of most of the topography between the Great Lakes and the Rocky Mountains has two distinct causes. In addition to the leveling and filling characteristics of the glaciers (which covered the area east of the Missouri River), the underlying bedrock surface itself is comparatively level. Under the ground layer of glacial or windblown materials are strata of flat-lying sedimentary rocks, often alternating in a sequence such as limestone/shale/sandstone/limestone. Beneath these rocks, in turn, is a massive slab of granitic rock, the Precambrian Shield, which is the core of the North American continent.

The Shield contains valuable deposits of metallic minerals which are accessible for mining in Canada and northern Minnesota, Wisconsin, and Michigan; but the Shield is buried thousands of feet below sedimentary limestones, shales, and sandstones in the Middle West. In a few places, such as the Black Hills of South Dakota or the St. Francois Mountains of Missouri, the ancient Shield rocks have punched upward through the newer sedimentary rocks, and in those isolated sites metal mining is an important economic activity. But these minerals are too deeply buried in much of the Middle West and Great Plains to be economically attractive.

Sand Hills of Nebraska

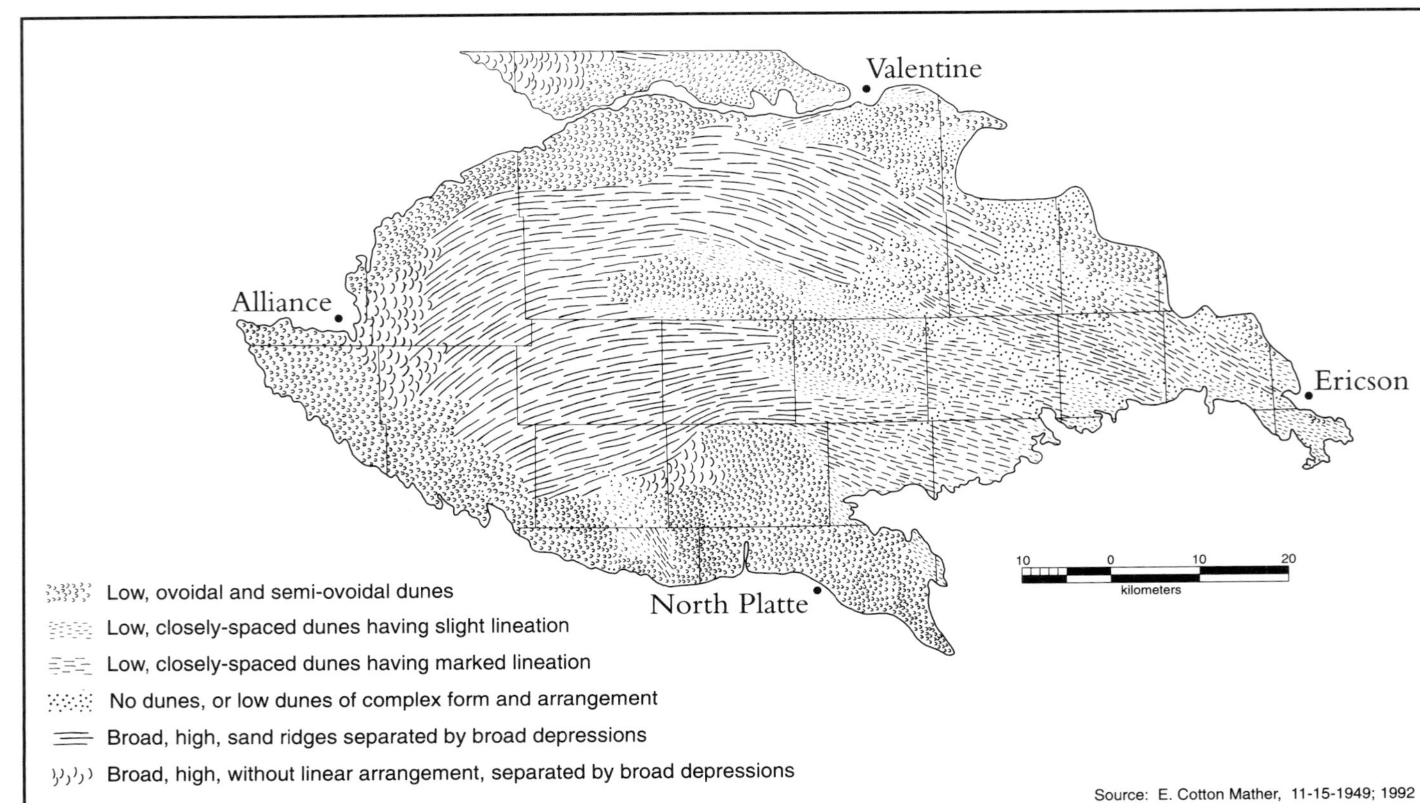

Layers of newer sedimentary rock superposed on the ancient Shield rocks were deposited in ocean basins or on continental margins millions of years ago. Some of these long-covered sediments contain valuable deposits of coal, gas, and oil. Central Michigan, most of Illinois, parts of Indiana and southern Iowa, and thousands of square miles in Wyoming, Montana, and North Dakota were so favored by the geologic past. The oil fields of the interior United States are not as productive nor are their reserves as large as those of Texas or Alberta, but the precious hydrocarbons have helped stimulate local economies in scattered locations. Coal is the most abundant hydrocarbon resource, and subbituminous coal reserves in the western Great Plains now are appreciated for their great economic value.

The United States, like all countries, is increasingly concerned with atmospheric pollution problems such as acid rain and the depletion of the ozone layer. The western Great Plains coals have a comparatively low sulfur-dioxide content, and their clean-burning qualities, despite their somewhat lower heat-giving properties, have reoriented the supply patterns of steam-coal production (mostly for electric utility consumption) in the past twenty-five years. The shift has come partly at the expense of coal production in the Illinois Basin. Both Indiana and Illinois have comparatively high-sulfur coals.

SLOPE AND DRAINAGE

The Middle West and Great Plains rest atop a stable platform. The rock layers underlying the surface have not been deformed; no ranges of mountains have been thrust upward by the tectonic forces of the Earth. This condition characterizes nearly the entire mid-continent regions from the Appalachians and their associated plateaus on the east to the Rocky Mountains on the west.

The relatively flat land surface between these fringing mountains is the drainage basin of the Mississippi River. With an area of more than 1.24 million square miles (3.2 million square kilometers), the Mississippi River basin is North America's largest, and in the world its peak flow is exceeded only by that of the Amazon and Yenesei rivers in South America and Asia. Rain and snow falling from the crest of the Rockies in the west to the summits of the Appalachians in the east eventually form rivers that flow across the mid-continent: the Ohio, Wabash, Kentucky, Illinois, Missouri, Cumberland, Minnesota, Platte, Wisconsin, Arkansas, Tennessee, and dozens of others. All of their water, in turn, eventually reaches the Mississippi. Beginning at the low-lying drainage divide with Lake Michigan, which occurs a few miles west of downtown Chicago, and continuing west to the peaks of the Rocky Mountains beyond Denver, the transcontinental traveler is within the massive Mississippi River drainage basin the entire distance.

Although the surface of the basin is relatively flat, elevation quickly increases to the west. Chicago, on Lake Michigan, and Davenport, Iowa, on the Mississippi, are nearly identical in elevation: 596 feet and 589 feet, respectively. West of there the regional slope is at first imperceptible, far less significant than the local variation in topography produced by river valleys. Iowa slopes an average of only sixteen inches per mile. This pattern changes at the edge of the Great Plains in eastern Nebraska, where the transition to newer, additional layers of sedimentary rock and thick surface deposits begins. The gradient averages six feet per mile from the Missouri River at Sioux City to the eastern edge of the Great Plains, eight feet per mile across the Nebraska Sand Hills, and then sixteen feet per mile across eastern Colorado before the approach to the Rocky Mountains begins. Denver, which officially lists its elevation at 5,280 feet (to match its claim as the "Mile-High City") is actually in a geologic basin at the edge of the Great Plains.

HUMAN OCCUPANCE

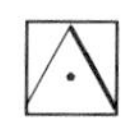

THE EARLY PERIOD

The St. Lawrence River–Mississippi River drainage divide at Chicago suggests one reason for the city's importance. It was once a portage between the two great systems of water navigation in the mid-continent. Lakes Superior, Michigan, Huron, Erie, and Ontario, plus the many dozens of relatively short, tributary streams flowing into them, are all part of the St. Lawrence River (Great Lakes) drainage. The early French settlers of Québec had to make only a single portage, somewhere around the southern rim of either Lake Erie or Lake Michigan, to reach a south-flowing stream they could follow uninterrupted all the way to New Orleans, Louisiana.

The earliest habitations of Euro-Americans in the Middle West were constructed by the French traders, missionaries, and soldiers who sought to fortify their sparsely settled but far-flung St. Lawrence-to-Gulf domain with a series of outposts. One of the oldest of these was Ft. Miami, constructed at St. Joseph, Michigan, on Lake Michigan, in 1679. The French built posts at Vincennes and near present-day Lafayette, Indiana, fifty years before Daniel Boone (b. 1734; d. 1820), who is often thought of as the first trans-Appalachian pioneer, took up residence as far west as Kentucky.

Early occupance of the Middle West and the Great Plains by Euro-Americans is celebrated today in countless communities across the land. Historic markers and buildings associated with the first settler, the first child born west of wherever, the first congregation of the whatever church, and similar firsts suggest that white settlers of Anglo-American heritage were truly "the first." In fact, this is rarely true. The frontier history of the United States is often interpreted as a series of firsts because, over most of the interior of the continent, it was the nineteenth-century settler who established the civilization that has thus far endured.

No one knows who first lived at the mouth of the Illinois River, or who first offered prayer west of the Mississippi, or who organized the first government of consenting citizens west of the Missouri. The archaeological record is silent on such matters, unyielding as to the particulars of persons and dates. The firsts we celebrate instead are those first recorded in deeds, records, or other artifacts of the comparatively recent (less than two centuries old) settlement by people whose ancestors came from Europe.

The "historic West" that the cross-country traveler learns something about from the roadside monuments includes little more than 2 percent of the time that humans have inhabited the region. Thus, to speak even of the early French outposts as "old" is to ignore about 97 percent of the human past; to start the clock with a buckskin frontiersman like Daniel Boone ignores 99 percent; and to assume that history began with permanent, white agricultural settlers from the East can be likened to describing a twenty-four-hour day by focusing on the part beginning eight minutes before midnight! The preceding twenty-three hours and fifty-two minutes, in conventional terminology, is known as "prehistory," the period preceding written records. Comparatively little is known about these people who are referred to in broadly descriptive terminology pertaining to their use of certain kinds of tools, structures, and foodways. We often call them "Indians" or "Native Americans," both terms reflecting errors (on the part of Christopher Columbus, who thought he had discovered the Indies; and by the German cartographer Waldseemüller, who labeled the continent "America").

Nor is it correct to equate the Indians whom whites first recorded as living at a particular locality with the original inhabitants of the place. While the antiquity of human presence is well-established (dating from about 12,000 years before the present over most of interior North America), it is mobility, rather than permanent settlement, that characterized most prehistoric settlers. One group might occupy a site seasonally over a period of several hundred years only to be replaced by another. Furthermore, there is no simple and direct linkage between prehistoric occupation and the Indian tribes of the historic period. The Miamis, Shawnees, Chippewas, Sioux, and others descended in some manner from the prehistoric peoples, but the precise connections are not well known. Even the relatively modern Indian tribes changed settlement strategies, moving from the forested east toward the grassy west, as westward pressure of the whites' frontier began in the seventeenth century.

THE FRONTIER

When did the frontier of white settlement cease its "inexorable" westward spread? The conventional answer is 1890, the year the Superintendent of the U.S. Census declared that a frontier line no longer existed. That date was about twenty-five years too early, for not until the first decades of the twentieth century did grain farmers expand into the dry rangelands of the western Great Plains. But 1890 is a good perspective from which to examine the massive changes wrought by a European-derived culture which, by that time, had nearly extinguished the civilization that preceded it. The near demise of the North American bison took place around 1890, the year U.S. soldiers massacred the Sioux at Wounded Knee in western South Dakota, often taken as the low ebb in American Indian history. Both the bison and the Indians had been isolated, herded into what remained of the interior grassland domain not yet put to the plow.

Treaty after broken treaty, the Indians were granted land "in perpetuity" a little farther west than they had lived before. Completion of the transcontinental Union Pacific Railroad in 1869 severed the Great Plains into two domains, north and south of the ribbon of iron rails that had been laid up the Platte River Valley and across the Great Divide Basin of Wyoming. Discovery of gold in the Black Hills in the 1870s produced another shrinkage of territory. Intertribal warfare, disease, and a diminishing supply of once-abundant food resources led to further declines. What remained of the open grasslands, where the bison and the Plains Indians roamed free, was subject to ever-increasing pressures, whether by cattlemen, gold-seekers, townsite boomers, or railroaders.

The nineteenth-century history of the Plains is one of a diminishing land mass for the Indians and the bison because of white encroachment from the east. California was already well settled in 1890, mainly by immigrants from the eastern states, but the dry grasslands of the Great Plains had been skipped over. Thus, it was in the Plains—in the middle of the continent—that the frontier was said to have ceased moving west, a decade before the twentieth century began.

The most frequently invoked model for interpreting this history was provided by American historian Frederick Jackson Turner (b. 1861; d. 1932). "Stand at Cumberland Gap," Turner wrote in 1894, "and watch the procession of civilization, marching single file—the buffalo following the trail to the salt springs, the Indian, the fur-trader and hunter, the cattle raiser, the pioneer farmer—and the frontier has passed by. Stand at South Pass in the Rockies a century later and see the same procession with wider intervals between."

Turner's chronology was correct, but his interpretation of the various waves of new settlers as a series of progressive steps in civilization has long been discredited. In 1894 Turner could only speculate what the next stages would bring. At that time, evidence suggested that urban industrial growth would follow the pioneer farmer in the West, just as it had in the East. Missing from Turner's scheme was any notion of regional economic interdependence, nor did he envision then that the stages were reversible, that once a particular civilization had been realized it might not endure.

A cross-country traveler of the early 1890s would have seen the logic behind Turner's scheme, though. Chicago was the economic capital of mid-America. West from there, for hundreds of miles across Illinois and Iowa, were the farms and small towns of the agricultural Middle West. Beyond farm country, in central Nebraska, civilization was still that of the cattleman, towns were few, and the land seemed not to have absorbed much of humankind's impact. A westward journey thus involved a trip through history, down through Turner's successive layers of human society. To go down this gradient of population density was to move away from complexity toward simplicity, to leave busy places for quiet ones.

CHICAGO

Chicago was then a teeming industrial city still assimilating the waves of European immigrants who provided the muscle for its busy factories. The city's fabric was defined by its radiating lines of railroad tracks and canals along which new industrial districts were expanding. Devastated by fire in 1871, Chicago had rebuilt quickly and in new ways. Architects were drawn to the city to design the new structures of brick, stone, and steel that replaced the old wooden buildings lost in the fire, a precedent for Chicago's future role as a center of innovative urban architecture. New annexations expanded the city's area fourfold by 1889.

Along the Chicago River on the north side of its central business district were the grain elevators and lumber yards that received wheat and saw timber from Wisconsin and the Upper Great Lakes region. On the south side lay the sprawling Chicago Union Stockyards, opened in 1865, which by the end of the nineteenth century were receiving nearly four million cattle per year from the railroads that entered the city from Iowa, Kansas, and Nebraska. Three fourths of the livestock were reloaded at the stockyards for shipment to eastern cities, the rest slaughtered in gargantuan packinghouses that lined the stockyards district. Chicago's central location drew trade from the east as well as the west.

One example of heavy industry was the Pullman Company's railroad-car manufacturing works south of the city. Pullman, a planned community founded in 1880 that even included farms to produce food for its inhabitants, was the scene of violence in 1893 when workers struck to protest wage cuts and the company's paternalism. Industrial community experiments like Pullman were not repeated. As manufacturing industries grew in importance in the Middle West, they were incorporated into existing urban centers or sited in new communities in which the industrial corporation was determined to play a minor role. Gary, Indiana, still nothing but swamps and sand dunes in 1890, would become the new model for the Middlewestern industrial city.

Chicago counted more than one million inhabitants for the first time in 1890 and its population grew steadily thereafter until growth turned to decline in the post–World War II period. Chicago's city population began to shrink when suburban growth attracted central-city dwellers outward to more than five dozen bedroom or satellite communities which, if they existed at all in 1890, had been small towns along one of the radiating main-line railroads.

Most of the spokes around Chicago's railroad hub were put in place between 1850 and 1870. Ten trunk lines ran to the east; three went to Michigan and seven crossed Indiana, linking Detroit, New York City, Pittsburgh, Cincinnati, Louisville, and Nashville to Chicago. Five main lines radiated southward toward St. Louis and New Orleans; six were aimed at Kansas City, Omaha, Denver, and the Far West; another five lines connected Chicago to Minneapolis-St. Paul, the Northwest, or Upper Michigan.

As the railroads approached Chicago they converged on no fewer than six major terminals in the city's "Loop" district. Until the 1970s, when service was reduced to a level that allowed all passenger trains to use a single Chicago terminal, the transcontinental traveler was obliged to detrain and cross the Loop, usually by taxi, to continue the journey. Truly like the hub of a wheel, the spokes of Chicago's radiating railroads made the city the center of the nation, the meeting point of lines that went to, not through, the city.

Today's urban/suburban landscape, which stretches an hour by expressway west of downtown Chicago, is a stage of civilization that Frederick Jackson Turner could not have envisioned. A Denver-bound traveler in 1890 would have been in open country by the time the present-day limits of Chicago were reached, just minutes from downtown. Now it is misleading to speak even of suburban development because the metropolitan area itself has decentralized, dispersed away from its former single focus, to the extent that the old central business district is just one node of metropolitan activity. Farmlands near the city have been developed for residences, then for shopping centers, now for office parks in a series of stages reminiscent of Turner's white hunter, cattleman, and farmer.

THE CORN BELT

In 1890 the dominant crop grown on the farms between Chicago and the Platte River valley of Nebraska was corn. In the 1990s, corn is still the farmers' main crop. The Corn Belt, a massive agricultural region stretching from Ohio to Nebraska, traditionally focused on raising corn for fattening hogs and cattle. With the construction of railroads in the mid-nineteenth century, parts of the central Corn Belt came to specialize in cash-grain production (off-farm sale of grain instead of on-farm feeding of animals). Chicago became a market for corn as well as for hogs and cattle. In fact, the Corn Belt probably was the single greatest factor in Chicago's industrial and financial growth. The city purchased and processed the region's grain and meat, built its farm machinery, and controlled its financial life. What Minneapolis-St. Paul, Kansas City, and Winnipeg were to North America's wheat regions, Chicago was to the Corn Belt.

The Corn Belt's Heartland location and its family-operated grain–livestock farms have led many to see it as the most typical of rural settings, while unglamorous, workaday Chicago, home to bootleggers and politicians, has been the American big-city prototype. Rain, sunshine, and fertile soils made the corn grow; industrious citizens, on the farm and in the city, made the landscape prosper. Hundreds of scaled-down, would-be Chicagos made their appearance along the tracks radiating away from the southern end of Lake Michigan. Each of them was a trade-center town that shipped the agricultural surplus of the surrounding farms, loaned farmers the money needed for the next year's crop, and sold them goods they did not produce themselves. Corn, hogs, mortgages, and orders for goods went up the line to Chicago; money, credit, goods, and enticements for more of the same came back down.

TOWNS, TRACKS, AND RIVERS

Along the tracks were the telegraph and telephone lines that communicated the transactions. They, too, focused on downtown Chicago, especially on the Board of Trade, where offers to buy or sell December yellow corn accumulated in great stacks, like the commodities themselves, every day. Night trains radiating from the city dropped off bundles of next-morning's *Chicago Tribune* in hundreds of outlying centers, binding the hinterlands to the city ever more tightly. To speak of the Middlewestern heartland as a settlement system is thus not to claim too much in the way of internal organization. The system that tied the outlying country to the city was most evident in the state of Illinois, but it organized Iowans and others as well. Even Minneapolis-St. Paul, Omaha, and Kansas City, the second-order metropolises of the mid-continent and important centers of banking and agricultural processing in their own right, were all about 450 miles from Chicago, well positioned to do business with the city in a day.

The Mississippi River and its many south-flowing tributaries had provided the network for a very different settlement system before the 1850s. St. Louis was the major metropolis of the mid-continent in the era of river commerce, while Chicago's swampy site on the drainage divide between the Mississippi and Great Lakes was far off-center, even irrelevant. The early urban centers of the Middle West were the river towns—Davenport and Keokuk in Iowa; Peoria, Oquawka, Galena, and Beardstown in Illinois. They traded mainly with St. Louis because that city—near the confluence of the Mississippi with the Missouri and the Illinois rivers, and a short distance upstream from the mouth of the Ohio—commanded centrality of position in the river network.

Railroads ran primarily east and west, the direction in which the nation was expanding, while the rivers flowed from north to south. A classic confrontation between the two took place in 1856, when the first railroad bridge across the Mississippi at Rock Island, Illinois, was "accidentally" rammed by a riverboat and destroyed in the ensuing conflagration. The boat's owners sued for damages, claiming that the bridge across the river constituted a material obstruction to river traffic. The railroad hired a young Illinois lawyer, Abraham Lincoln (b. 1809; d. 1865), to present its case. Lincoln's reasoned presentation of the facts in the case, long since part of American folklore, led to the railroad's victory. From that time onward, the river towns dwindled in importance, their logistical advantage erased by a new form of transportation that bridged high over the river and its once-busy levees. Commercial patterns of the mid-continent were reoriented east and west along the alignment of Chicago's railroads, instead of south down the Mississippi to St. Louis and New Orleans.

By 1890 Illinois had 10,000 miles of railroad, the most of any state; Iowa and Kansas were second and third in rank, respectively. Similar developments in southern Wisconsin, eastern Nebraska, and Indiana also were undertaken by railroads with Chicago destinations. The Middle West, especially the Corn Belt, eventually became laced with railroads, about seven miles of track for every 1,400 inhabitants. Seven miles represented the approximate distance between towns along the track; the 1,400 inhabitants were divided, about two-to-one in 1890, between farms and towns in the rural areas. While not every farmer had a direct link to metropolis, every town did and many had more than one railroad line.

These were ingredients in a recipe for standardization: thousands of farmers, all selling the same kinds of produce; hundreds of towns, each selling the same mix of goods. Large towns offered more variety than small ones; thus, even within small regions of the Corn Belt a hierarchy of alternatives and options developed. Every place attempted to mimic life in the larger, nearby centers, yet all attempted to instill in their citizens the unique qualities of their own town or locality. Loyalty, whether in politics or trading allegiance, became a part of place-consciousness.

THE LAND SYSTEM

Another repetitive motif is the landscape-checkerboard of 160-acre (65-hectare) farms resulting from the rectilinear, township-and-range survey system that subdivided the public lands. Often referred to as homesteads—after the Homestead Act of 1862 which allowed persons to obtain 160 acres of public land free by proving residency on the land of five years—these building-block units of the rural landscape had been accumulating in land-office records for years before the Homestead Act was passed. "Free land in the West," a rallying cry for many in the mid-nineteenth century, was a boon to settlement on the fringes of the Middle West not yet reached by the westward-moving tide of land-seekers.

Turner's waves of settlers can be seen in the homestead entry statistics. Illinois, well settled by 1862, eventually recorded only seventy-four homesteads in the entire state. Moving west, Iowa had 8,851; Nebraska, 104,260; and Colorado, 107,618. As the frontier reached farther toward the dry plains, land laws were liberalized to allow the farm unit to become larger. The size of homestead entry progressed from 160 to 320 and, finally, to 640 acres (one square mile), although the law lagged well behind the need to increase farm size. Thousands of families were in jeopardy on the dry western Plains, where even a square mile of land was not enough.

The checkerboard-grid of farms has been criticized for its relentless subjugation of all lands, regardless of quality, under a simplistic geometric plan. The American land grid, first established in Ohio with the Land Ordinance of 1785, left no special places for towns or other points of social and economic congregation. It ignored community by dispersing population nearly to its theoretical maximum, putting people close to the land instead of close to each other. Providing services was costly in such a dispersed population and the result, in many areas, was a diminished quality of life. Electricity and telephones came to the farm later than they did to town. The very problem of "getting to town," unknown in the traditional agricultural villages of Europe, where farm and town coalesced, was a hallmark of rural settlement in the largest, richest agricultural region in the United States.

Minor exceptions to the rule suggest what an alternative might have looked like. A group of German pietists who had come to the United States and founded the Ebenezer community near Buffalo, New York, in the 1820s later moved to Iowa. They founded five European-style agricultural villages—the Amanas—along the Iowa River southwest of Cedar Rapids. A communistic society, the Amana colonies nonetheless were a capitalistic success. Part of the profits made from the sale of their lands at Buffalo was invested in manufacturing enterprises at the Amanas. The rural Middle West might have looked more like the Amana colonies had there been more groups with similar goals, but communitarian settlements were not attractive to most people, then or now. The Amanas, like the industrial city of Pullman, were isolated attempts to improve the lot of the farmer or city-dweller, but they did not attract support from the masses.

CHANGING TIMES

The so-called Jeffersonian ideal, a rural landscape populated by freeholders, was the intent of those who framed the public-land laws. But land, whether in the American democracy or the old European monarchies, did not remain in the same-sized parcels forever. New generations replaced the old and they subdivided or sold the land, treating it less as a birthright than as an asset that could be sold or borrowed against. Whether from land sales, mortgage foreclosures, or tax forfeiture, the land the settlers had alienated from the public domain in the nineteenth century slowly began to accumulate in the hands of bankers, insurance companies, and other absentee holders in the twentieth.

Although farm tenancy sometimes is portrayed as accompanying agriculture on marginal land, the reverse has more often been true in the Middle West. Good land, not poor, was most attractive to those who wished to accumulate large acreages, subdivide them, and rent farms to tenants. Those who loaned money similarly were interested in the quality of the land offered as collateral. The best croplands in the Corn Belt were especially valuable because of the improvements that had been made to them.

Much of this high-quality land was created in late nineteenth- and twentieth-century Iowa and Illinois by digging ditches and laying thousands of miles of underground drainage lines on the wet prairies. Soils that formed this poorly drained, swell-and-swale topography where glaciers had remained for extended periods had high natural fertility but were unusable for crops without some means of draining off the excess water. Draining the lands cost money and hence their value increased with the additional investment. The most valuable lands became, simultaneously, those that produced the most bountiful harvests, had cost the most to bring into production, were most likely to be mortgaged, and were most likely worked by tenants.

Fewer than one in four Iowa farmers were tenants in 1880, but by 1930 nearly half of them were. The pioneer farmers in Turner's stage of civilization's advance had slid backward, to a semi-feudal condition. Agricultural depression in the 1920s was followed by a more severe, general depression and searing droughts in the 1930s. But the Corn Belt did survive. Recovery after World War II was followed by new patterns of agriculture. Tenancy decreased as farmers acquired small acreages, expanded them, and rented still more land. The introduction of hybrid corn more than doubled yields. Soybeans, practically unknown to American farmers before 1920, became an important cash crop by the 1940s. Continued mechanization, increased use of chemical fertilizers, and reliance on herbicides caused grain outputs to skyrocket.

The Corn Belt farm has been greatly affected. There are only two-fifths as many farmers in the Middle West today as there were even in 1950, although the acreage in farms has remained relatively constant. Today's farmer relies more on cash-grain sales than on livestock production. The feeding of beef cattle has moved westward into new, irrigated districts of corn and small-grain production on the Great Plains. Fewer farmers working larger acreages means fewer patrons for trade-center towns. People have disappeared as the region's agricultural output has increased.

The 60 percent decline in the number of Middlewestern farmers is far less than the almost 90 percent decline in the South and Northeast. About half the nation's employment in agriculture occurs in the Middle West today, up from one-third forty years ago. Yet, although agriculture is by far the largest user of land in the Middle West, it is not the major employer in any but the most rural areas. In fact, a slight majority of Middlewestern farm residents now have their primary employment off the farm. A major transition in the economic base of small towns has accompanied the shift away from reliance on agriculture. Many small towns and cities of the mid-continent now have significant manufacturing employment. Some of the industries are agriculture-related, but many are not. Today one can even find steel mills in some of the Corn Belt's small cities.

THE GREAT PLAINS

The major change in the landscape's appearance noticed by travelers as they cross the Heartland, whether in the 1890s or the 1990s, occurs at the eastern edge of the Great Plains. In 1890 the vast northern interior of Nebraska—the Sand Hills—was cattle country; and so it remains today. The Kincaid Act of 1904 opened the Sand Hills to homesteaders by allowing 640-acre homesteads, the larger land allotment being necessary because the area was suited mainly for cattle grazing. Livestock thrive on the grasses, but breaking the soil surface to plant crops is risky at best. Largely because of its unique physical nature, the Sand Hills has remained a livestock-ranching country.

Not so with the dry, short-grass plains of western Nebraska and eastern Colorado, however. They, too, were largely devoted to livestock range at the beginning of the twentieth century and were thought to be unsuited to farming, given a precipitation average of less than seventeen inches (forty-three centimeters) per year. During World War I, when the demand for American wheat reached record levels, farmers expanded westward into the driest lands of the Great Plains. Stockmen retreated in the face of this movement, the last westward push of the American pioneer farmer. The "big plowup" of the 1910s came to an abrupt halt when oversupply caused the price of wheat to plummet in the 1920s. During the drought of the 1930s, many abandoned their farms in this land of extreme risk where, in some counties, crops are left unharvested two years in five because of lack of moisture. But risk-taking wheat farmers have always returned to reoccupy the land. The Colorado/Nebraska High Plains remains today a major producer of wheat. Only the most marginal lands have been placed in federal grassland reserves designed to prevent cultivation of erosion-prone soils.

For years a heated debate has raged over how the Great Plains ought to be used. Some believe, along with geologist and explorer Major John Wesley Powell (b. 1834; d. 1902) more than a century ago, that most of the land should remain as grassland and be used only for pasture. Also like Powell, those who regard the Plains as deficient in moisture have espoused schemes for irrigating parts of it, especially valley bottoms along the major streams. Many miles of irrigation ditches have been constructed, but new irrigation developments, especially the center-pivot sprinkler methods, now surpass the importance of ditch irrigation.

The center-pivot system's ability to irrigate rolling, upland fields greatly expanded the area where irrigation could be attempted. Water pumped up from underlying bedrock aquifers has expanded crop farming just as it has provided a cushion against failure from drought. The world's demand for corn, especially, has stimulated a westward expansion of the Corn Belt into areas once suited only for dryland wheat. The old debate as to whether to use uplands for grazing or for crops has shifted to the question of whether to raise irrigated corn, dryland corn, or dryland wheat. Using such lands only for grazing is now out of the question, given their increase in value. The sweep of Middlewestern-style cornfields and livestock feedlots has been completed across much of the Plains.

DENVER

In 1890, when Chicago's population had reached the one million mark, Denver was only about one-tenth that size. Denver's growth and Chicago's comparative stagnation have reduced that ratio closer to six-to-one today, although Denver's city population has declined in recent years as well. If Chicago typifies the Middle-western industrial city, Denver deserves the Sunbelt category. Like other cities of the West and Southwest, Denver was a mining boom town and a cow town with rough saloons. Now it is a corporate headquarters city playing a major role in the era of electronic media. Denver is the sort of clean, good-life city to which Chicago's young generation moved after World War II. Chicago struggles to keep its industries, while Denver seems not even to have to try to get new ones.

Differences between the two reflect the length of time it took the population frontier to move west across the United States. Chicago's rusting industrial structures have some counterparts in Denver, but the latter city has become a major metropolis in a new age. The center of the nation's population is marching away from Chicago, veering both south and west toward the amenities of life which fewer people each day seem to find in the Middle West.

To travel east to west—from Chicago to Denver—is the proper way to perceive this historic change in the American land. And Denver is a fitting place to pause in a transcontinental journey because this city, which is of both the plain and the mountain, is more westward-looking all the time.

PART TWO

The Itinerary

Crossing the Heartland

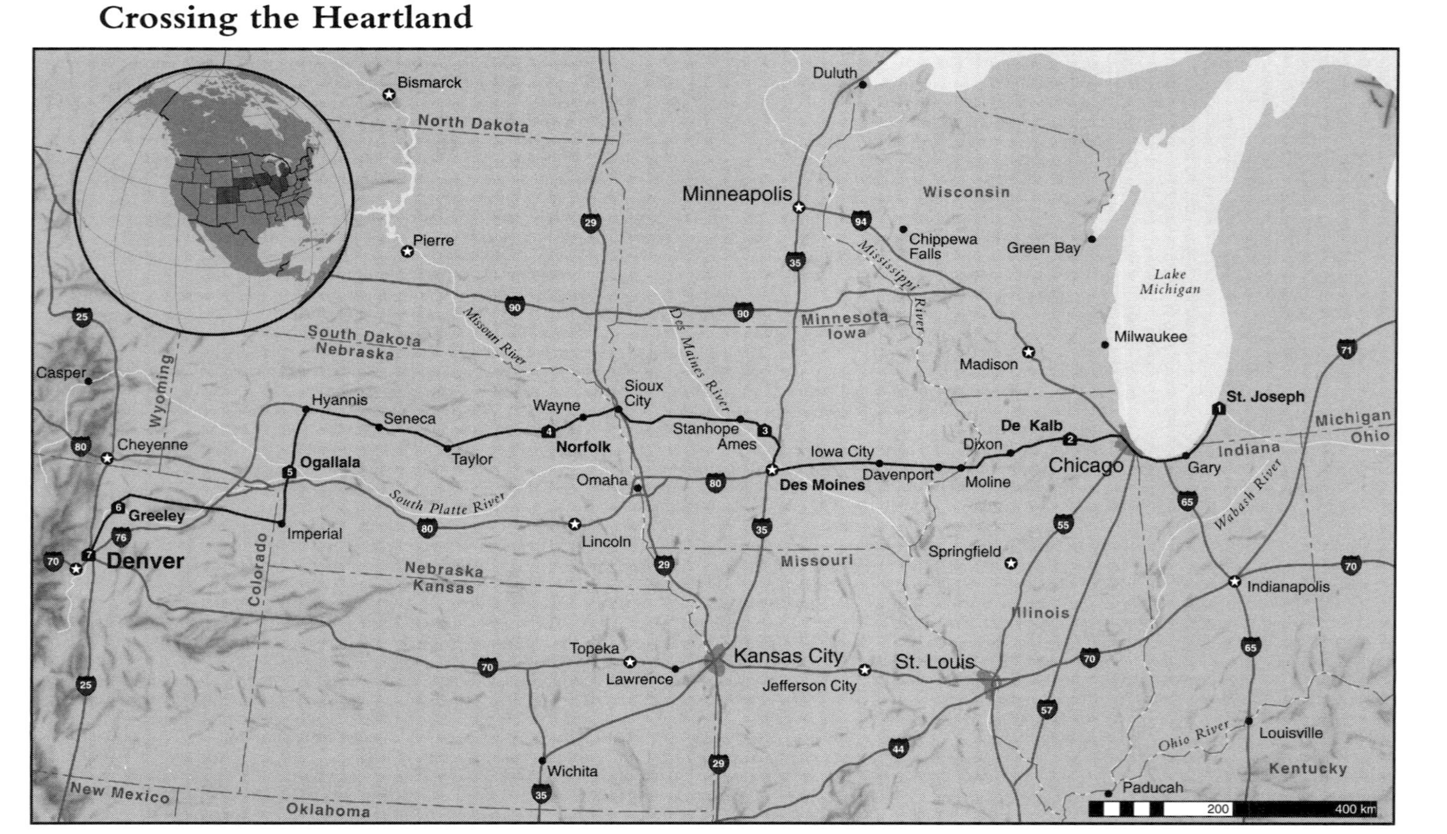

CROSSING THE HEARTLAND

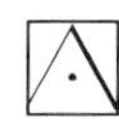

Our tour, "Crossing the Heartland," begins and ends in two great cities—Chicago and Denver—but along the way we shall also experience the wide open spaces of the land in between and the many other urban enclaves (even if they are small towns) whose importance to their region will become self-evident as we pass on through. Nature will abound—from the Indiana Dunes to the prairies to the Sand Hills to the foothills of the Southern Rockies. Culture will abound—from Chicago's Loop to the changing urban functions of Mississippi River towns to the Mom and Pop stores where one can quench one's thirst. This segment of the Transcontinental is through middle America, and visions of the Heartland no doubt abound in your mind prior to the first day's travel.

We shall do our best to get off the interstate to explore the city streets of metropolitan Chicago and the town streets of Main Street U.S.A., to get onto the back roads where one can actually see and experience the landscape. We will see a lot, and no doubt, when you are in Denver at journey's end, and are looking to the west into the Southern Rockies, you will not soon forget the passage across the Heartland.

St. Joseph, Michigan, to De Kalb, Illinois

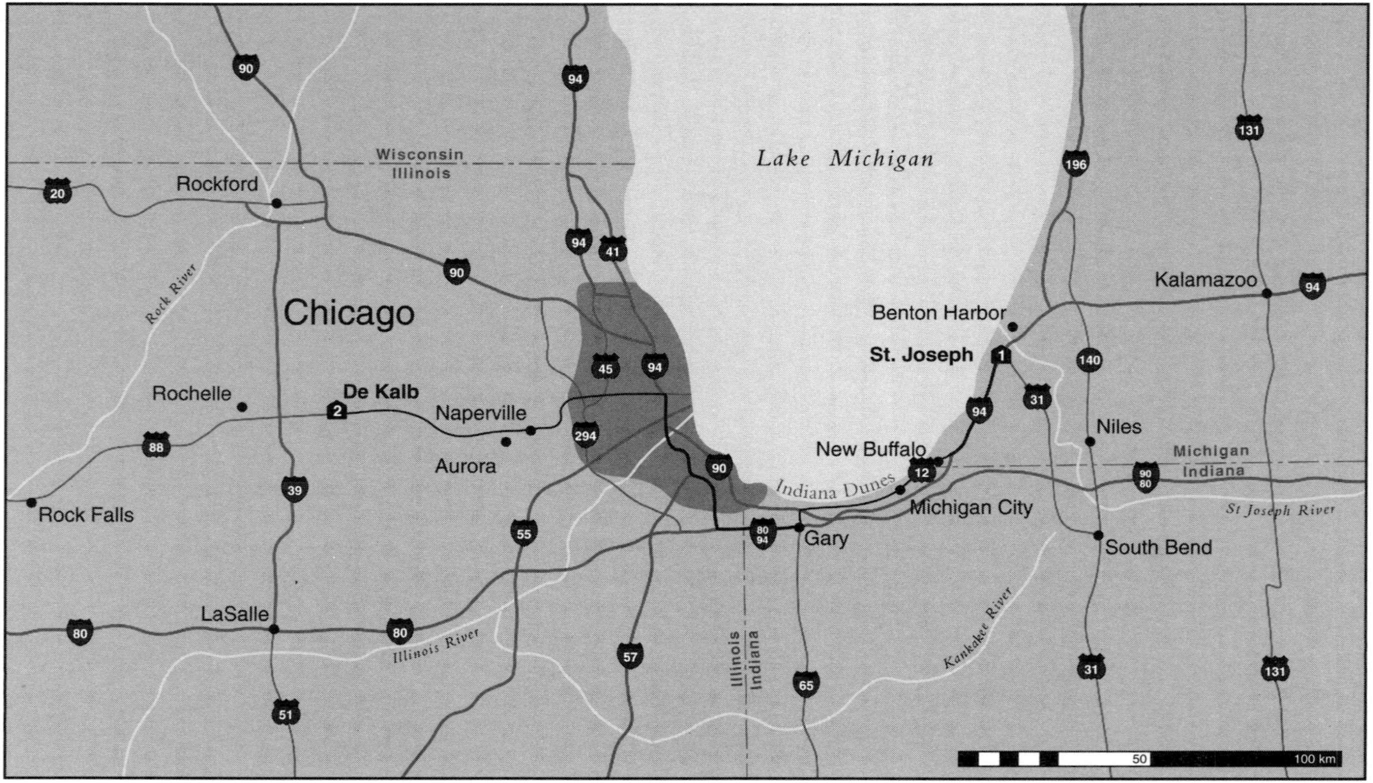

Day One

CHICAGO AND LAKE MICHIGAN
St. Joseph, Michigan, to DeKalb, Illinois, 165 miles

Crossing the Heartland begins by following the shorelines of Lake Michigan, via the Indiana Dunes, to the city of Chicago. This is the most industrialized section of the trip. Rural landscapes will be seen only in the opening and closing hours of the day's travel.

St. Joseph (population 9,215) begins a continuous strip of metropolitan areas that stretches around the southern end of Lake Michigan and into Wisconsin. Its site also marks one of the oldest Euro-American outposts of the Middle West. René Robert Cavalier (b. 1643; d. 1687), Sieur de La Salle, arrived here under sail in 1679. La Salle built Fort Miami at the mouth of the St. Joseph River (then called the Miami), a route the French thought would be important for future travel down the Mississippi River. They followed the St. Joseph upstream (southeast) to the vicinity of Niles, Michigan, where a short portage was made to the Kankakee River, a tributary of the Illinois River.

The other city at the mouth of the St. Joseph, *Benton Harbor* (population 12,820), was settled by Euro-Americans in 1834. Its sheltered harbor, like that at New Buffalo just to the southwest, once had a strategic value. By the 1830s, Detroit and southern Michigan were linked via overland wagon road to these ports on Lake Michigan which made for faster travel than the all-water route around Michigan's lower peninsula.

Both St. Joseph and Benton Harbor became manufacturing cities by the mid-nineteenth century. Today they illustrate divergent paths in the recent history of industrial cities. Benton Harbor's central business district is virtually deserted. Its storefronts either have been abandoned or recycled into some lower-value use than they had before. Investment capital has fled, the city has been plagued by political corruption, and its crime rate is high.

Across the river, on the high bluff overlooking the lake, is St. Joseph. It has been transformed into a downtown mall of stores and boutiques that attracts summertime tourists visiting the Lake Michigan beaches. St. Joseph's business streets, lined with historic buildings dating from the early decades of the twentieth century, have been preserved through adaptive reuse. Both cities have had to adjust to a decline in the regional industrial-economic base. The "decay and abandonment" alternative is here set clearly alongside the "mall and boutique" alternative. As important industrial cities, however, both are history.

St. Joseph to New Buffalo, Michigan, 32 miles, I-94

Business I-94 south joins I-94 to Chicago 6 miles south of St. Joseph. The highway traverses a lowland hardwood forest fringing sand dunes. The dunes along Lake Michigan's southeastern edge and the occasional appearance of apple orchards and grape vineyards are both related to geographic position. Westerly winds blowing across the lake provide the force necessary to mass the shoreline sands into large dunes. The prevailing westerly winds moderate the local climate, providing more marine effect on the southeastern side of the lake than elsewhere. The growing season is longer, and winter temperatures are less severe. "Lake effect" here means that snowfall is heavier, however.

The dunes bordering Lake Michigan were scarcely used before modern times. Too hilly and sandy for cultivation, they offered little to the early settlers. The highway passes the Bridgman Nuclear Generating Station and an active sand mine (both near the Bridgman exit from I-94).

New Buffalo, Michigan, to Gary, Indiana, 35 miles, Highway 12

Leave I-94 at New Buffalo (exit 4B) and follow U.S. Highway 12 west. *New Buffalo* (population 2,320) was briefly an important lakeport. The Michigan Central Railroad was completed from Detroit to New Buffalo when Chicago still lacked a railroad line to the east. Today pleasure boaters use New Buffalo's dredged harbor, and resort condominiums line the waterfront area. Chicagoans flock to these shores in summer to catch the cool lake breezes. Numerous restaurants and motel accommodations are found in this stretch.

Crossing the Michigan–Indiana state line, Highway 12 continues to follow the lake-border fringe of sand dunes. Land-use intensity increases approaching *Michigan City* (population 33,825), which has long been within daily railroad commuting distance of Chicago. Leaving Michigan City the highway passes the mammoth cooling towers and electric-power–generating station (coal-fired) of the Northern Indiana Public Service Commission.

One mile past the power plant enter the *Indiana Dunes National Lakeshore* at the sign indicating "Mt. Baldy." Parking is available for cars and tour buses. A trail leads from the parking lot into the dunes. Mt. Baldy is a spectacular example of a coastal blowout dune. Prevailing winds remove sand from the beach, while the fringing forest serves as a natural barrier against which the sand accumulates. The result is a massive ridge of sand, here swept up to a height of 125 feet. The lake side of the dune has a concave profile while the landward side is straight; both are steep. All of Indiana's dunes (as well as southeastern Michigan's) were formed by the prevailing westerly winds acting on the sandy lake margins, which is why the dune landscape extends no farther west than the southern tip of Lake Michigan.

The present-day outline of Lake Michigan suggests the general shape of a massive lobe of the continental ice sheet. Geologists have given the name *Lake Chicago* to the body of water formed from melting glaciers that lay approximately where southern Lake Michigan is now. A series of morainic ridges (representing glacial advance and deposition) and sand dunes (glacial retreat and melting) parallel the present lakeshore. The dunes were formed during successive water-level reductions in glacial Lake Chicago. The present dunes have formed within the past 10,000 years. Highway 12 follows the crest of the Calumet Dunes, formed 11,500 years ago when Lake Chicago was 40 feet higher than present-day Lake Michigan. In all, 4 sets of morainic ridges parallel the lake, most of them accompanied by some dune features. Farthest inland is the Valparaiso moraine, an abrupt line of hills approximately 5 miles to the south. The oldest landforms thus lie farthest from the lake. The dunes now forming rise abruptly from the present beach edge just as older dune formations did in past periods.

Return to Highway 12 and continue west. The highway here is shaded by an immense canopy of old-growth hardwood forest (beech, elm, maple, and oak). Sandier and more recently disturbed lands have a shrubby cover of sumac. Wetlands between the ridges of dunes support aquatic plant communities. The rich diversity of flora here caught the attention of University of Chicago botanist Henry C. Cowles (b. 1869; d. 1939) in the late 1890s. Cowles recognized that the first plants to colonize a dune surface ("pioneers") were succeeded by more nutrient-demanding species. The many dune environments aided Cowles in formulating his influential theory of vegetation succession.

Paralleling Highway 12 is the track of the Chicago, South Shore, and South Bend Railroad, the last surviving interurban electric railroad line. The cities of the American Manufacturing Belt once were linked in a network of these electrically powered lines. The "South Shore," part of utilities tycoon Samuel Insull's fallen empire of the 1920s, still shows his influence in the steel-girder trolley-wire supports lining the track. Now part of the northern Indiana mass transit district, the South Shore line remains a functional part of the metropolitan commuting system. Classic exam-

The steel industry in the Calumet region of Indiana is dominated by the large integrated steel manufacturing facility of Bethlehem Steel Corporation at Burns Harbor. Encroachment by industries like this one led to the authorization of the Indiana Dunes National Lakeshore in 1966 and its establishment in 1972. Photograph by John C. Hudson.

ples of historic station designs along the line include *Beverly Shores* (population 600), which serves summer and year-around homes in the Indiana Dunes. The new station at *Tremont Dunes* suggests the railroad's continued importance.

The landscape changes at the western edge of the Indiana Dunes National Lakeshore with an abrupt transition from forest to industry. In fact, the Lakeshore preserve was authorized in 1966 and established in 1972 largely as a response to the encroachment of industries into what remained of the dunes. This is the *Calumet Region* of Indiana, a sprawling complex of steel mills, refineries, and chemical plants that began to take shape early in the twentieth century and reached its greatest national and strategic significance during World War II. The sandy dunes bordered by swamps and drained by sluggish streams (such as the *Calumet River*) repulsed early settlers who went toward better lands beyond Chicago. This large tract of unused land, all of it fringing Lake Michigan and the access it offered to shippers of bulk raw materials, was left unclaimed until Chicago's industrial potential became known. The Calumet region went from unused land to heavy industry in a single step.

One of the nation's largest, fully integrated steel mills is operated by Bethlehem Steel at *Burns Harbor* (large entrance signs west of Tremont Dunes). Farther west on Highway 12 are entrances to the *Port of Indiana* and another steel complex, the Midwest Division of National Steel. The Port of Indiana handles a variety of incoming bulk cargoes; it also loads Great Lakes ships with Indiana corn, soybeans, and wheat, much of which is exported.

Gary, Indiana, to Chicago, Illinois, 30 miles, I-90

Take Interstate 90 (Indiana Toll Road) west at its junction with U.S. Highway 12. The new steel mills and ports just seen are replaced here by the older commercial and residential sections of *Gary,* Indiana (population 116,645). Created by a real estate affiliate of the United States Steel Company in 1907, Gary's gridiron of city blocks adjoined the blast furnaces and rolling mills where its inhabitants would work. A once-proud working-class city, Gary's fortunes began to slide in the 1950s. Today, employment in the steel mills has partially stabilized at a lower level, but, with insufficient replacement jobs for its citizens, Gary continues to decline.

The Gary Works of United States Steel is on the north (Lake Michigan) side of the highway in downtown Gary. Two miles beyond the main plant is the old office building of the American Bridge Company, a U.S. Steel subsidiary. The Ambridge Building, dating from 1910, displays typical company architecture of that period.

The toll road then passes south of *Whiting* (population 5,155), another company town, in this case the work of the Standard Oil Company which built massive refineries here that connect by rail and pipeline to the Texas and Gulf oil fields. Just west of the toll plaza is the office of AMAIZO, a corn-products processor whose Chicago-area location typifies this industry which depends heavily on the eastern Illinois and western Indiana corn crop. (Watch for signs pointing to Phil Smidt's restaurant in Hammond, Indiana, a 1950s-style fish house and local tradition specializing in frogs' legs as well as lake fish.)

CHICAGO'S SOUTH SIDE

The Indiana–Illinois state line here is identical with the city limits of *Chicago* (population 2,783,725), now the third largest city in the United States (after New York City and Los Angeles). Just east of the state line the Indiana Toll Road becomes the *Chicago Skyway* which bridges high over the Calumet River and offers a panoramic view of the heavy industrial sector of Chicago. The Calumet River has been dredged for deep-draft vessels which pass beneath the bridge to reach the port of Chicago/Lake Calumet. Leave the Skyway at the Stony Island Avenue exit.

Proceed north on Stony Island Avenue. Here is one segment of Chicago's south-side black community. At 73rd and Stony Island are the Mohammed University of Islam and Mosque Maryam, headquarters for Chicago's controversial Muslim leader, Minister Louis Farrakhan. Ahead (north) on Stony Island lies *Jackson Park.* Jackson Park was originally designed by Frederick Law Olmsted and Calvert Vaux, but John C. Olmsted did much of the design that was implemented after the Columbian Exposition. To reach the *Museum of Science and Industry,* follow Cornell Drive at its bend east from Stony Island north of 57th Street. The museum, whose main building was the Fine Arts Pavilion of Chicago's 1893 Columbian Exposition, is one of Chicago's major cultural attractions and it is the most visited museum in the United States. Parking is available on the north side of the museum.

A panoramic view of the *University of Chicago* campus can be had by following 59th Street west from Stony Island Avenue. The

SIDE TRIP TO PULLMAN AND THE PORT OF CHICAGO

Exit the Skyway at Stony Island Avenue and turn back south on Stony Island which soon becomes Interstate 94. Less than a mile south take the 111th Street exit off I-94 and travel west. Turn south at 111th and Langley to reach the Pullman historic district. The Florence Hotel and other buildings here remain from the community that railroad-car builder George M. Pullman (b. 1831; d. 1897) founded in the early 1880s. Laid out by architect Solan Beman on 3,500 acres then well outside the Chicago city limits, Pullman was thought to be a model community that overcame the evils of the industrial city. The depression of 1893 led to layoffs at the Pullman car works and the layoffs were followed by a strike. Violence came when President Grover Cleveland (b. 1837; d. 1908) ordered in federal troops. The Illinois Supreme Court found Pullman's paternalistic social design for a community to be unconstitutional and the workers' homes were sold to them after 1898.

Return to I-94 and travel south to the 130th Street exit; follow 130th Street east to Torrance Avenue to view the Port of Chicago. Then turn north on Torrance, past the Lake Calumet harbor-area facilities and complete the loop by turning west at 103rd and Torrance to return to Stony Island Avenue.

open "Midway" between 59th and 60th Streets also dates from the 1893 Columbian Exposition. The University of Chicago, founded in 1892, stretches several blocks north of 59th between Blackstone and Cottage Grove avenues. Campus architecture copies that of Cambridge University in England. Its neighborhood, Hyde Park,

SIDE TRIP TO THE SITE OF THE FORMER CHICAGO UNION STOCKYARDS

Follow 59th Street west to the Dan Ryan Expressway (Interstate 90–94). Take the expressway north to the 43rd Street exit (four exits north of 59th). Follow 43rd Street about ten blocks west to Halstead. The old International Livestock Exposition amphitheater is at 43rd and Halstead. Turn right and travel two blocks north on Halstead, then one block left on Exchange Avenue to reach the stockyards site.

All that remains of the Union Stockyards is the original stone entrance gate and some rusting rails that once were part of the loop of railroad tracks that encircled the yards and served more than thirty meat-packing houses. Opened in 1865, these were the world's largest stockyards, employing more than 30,000 people at their peak in 1920. But, by the 1950s, the meat-packing industry began to move west and cattle and hogs more often arrived at the stockyards in trucks than in railroad cars. By 1971 the business had vanished and the yards were closed. Today, light manufacturing and warehouses occupy part of the site. The nearby International Amphitheater, home to countless livestock expositions, hosted the Democratic national conventions of 1952 and 1968 (the latter infamous for its confrontation between Vietnam War protesters and Mayor Richard J. Daley's Chicago police).

Return to the Dan Ryan Expressway and proceed north, following signs for "Lakeshore Drive" to reach Chicago's Grant Park/lakefront area.

has a majority white population but is bordered by areas that are more than 90 percent black.

The predominantly black section of Chicago now stretches more than 15 miles from the city's southern edge to the downtown area and from there extends straight west another 8 miles. Black migration, especially from the Southern states, increased after World War I when the demand for industrial labor in Chicago could no longer be met by European immigrants alone. Blacks now make up about two-fifths of the city's population, although as recently as 1960 their proportion was only slight above one-fifth. The black population is growing comparatively slowly today, however, while the sharp increase in immigration of Spanish-speakers after 1970 plus "white flight" to the suburbs has made Chicago a white/black/Hispanic city. In recent years a large influx of Asians has also changed the ethnic composition.

Chicago may be the most segregated city in the United States. Most blacks live in neighborhoods that include no more than 10 percent others and in many places the racial/neighborhood cleavage is severe. Western Avenue on the city's west side is one such division which separates almost entirely black area on the east from white neighborhoods with a strong European-ethnic flavor on the west. Any route the traveler takes across the South Side of Chicago (and some other sections as well) will be within view of one or more high-rise Chicago Housing Authority projects for low-income residents. Some of these "projects," along with hundreds of city blocks of run-down apartment buildings, are among the most lethal environments in the United States as measured in statistics such as murders per day. Outsiders should not linger here.

From the Museum of Science and Industry or the University of Chicago campus automobiles (trucks and buses not allowed) take Lakeshore Drive (U.S. Highway 41) north to the center, "the Loop."

DOWNTOWN CHICAGO

One of the world's most impressive skyscraper panoramas, downtown Chicago is best viewed across Grant Park from Lakeshore Drive. Another vantage is obtained from the observation windows atop the *Sears Tower* (the world's tallest building) at West Adams Street and Wacker Drive or from atop the *Prudential Building* (Michigan Avenue at Randolph Street), which gives a better view to the south.

The traditional heart of Chicago, *the Loop,* is defined literally by a loop of elevated railroad tracks, known as the "El," that straddle Wabash (east), Van Buren (south), Wells (west), and Lake (north) streets. The Loop includes part of the city's original six- by ten-block grid which Illinois and Michigan Canal commissioners laid out at the junction of the north and south branches of the Chicago River in 1830. The city's population had grown to 30,000 by 1850. The disastrous fire of 1871 devastated the area between the two branches of the Chicago River and Lake Michigan. The need to rebuild encouraged innovation, and some of the world's first steel-skeleton skyscrapers were built in Chicago by the early 1880s. The Loop area retained its role as the commercial and financial heart of the city, while new industrial districts extended in all directions along the railway lines entering the city.

Although the loop of elevated railway tracks still functions as an important part of Chicago's transportation system, areas just outside the Loop have seen most of the architectural and commercial developments of recent decades. *State Street,* once the most fashionable shopping district, has given way to the so-called *Magnificent Mile* of Michigan Avenue north of the Chicago River. Twin communications spires identify the John Hancock Tower in this neighborhood, which also includes Saks Fifth Avenue, Neiman Marcus, and Water Tower Place department stores.

New skyscrapers west of the Loop, especially along Wacker Drive, have accompanied the westward expansion of Chicago's financial district, which focuses on several major banks and commodities exchanges. The *Chicago Board of Trade* (founded in 1848) has a Visitors Center on the fifth floor of its building at 141 West Jackson Boulevard. From the glass-enclosed observation area (flash pictures are prohibited) one may view the hectic pace of trading activities in this, the world's oldest and largest futures exchange.

Traditional Chicago attractions, such as the mammoth Marshall Field department store and Palmer House hotel on State Street and the Art Institute on Michigan Avenue continue to define points of interest along the eastern edge of downtown. While new skyscraper construction has moved north and west of the Loop, it remains conspicuously absent to the south.

Daytime travel by foot or by subway and elevated train (fare $1.50, as of this writing) is safe and the Loop's near-perfect grid of blocks makes way-finding easy. Because the area is primarily a workplace, its nighttime population, outside of restaurant and theater areas, is comparatively sparse.

Chicago to De Kalb, Illinois, 68 miles, I-290 and I-88

Leaving downtown Chicago, travel west via the Eisenhower Expressway (I-290) which is accessed from Congress Parkway just south of the Loop. This is one of the city's busiest expressways, and travelers should expect delays during morning and afternoon rush hours. The Expressway leaves Chicago's city limits at Austin Boulevard (6000 West) and traverses the old suburbs of *Oak Park* (population 53,600), *Maywood* (population 27,100), and *Belwood* (population 20,200) before crossing the Tri-State Tollway (I-294) at *Elmhurst* (population 42,100).

At the junction of I-290 and I-294 continue west and enter the East–West Tollway (I-88). About a mile before this junction the highway climbs a slight rise out of the glacial Lake Chicago lowland and crosses the hillier Tinley moraine which constitutes the old lakebed's western edge. A series of low rises to the west as far as *Naperville* (population 85,400) marks the position of earlier advances of the Lake Michigan glacial lobe.

The growing cities of *Oak Brook* (population 9,180) and *Downers Grove* (population 46,860) typify the modern suburb where workplaces and shopping centers have been added to the traditional "bedroom community." The East–West Tollway, also dubbed the *Illinois Research and Development Corridor,* traverses one of the new suburban corporate headquarters axes radiating from the city. More than a dozen companies, including Amoco, Bell Labs,

The Corn Belt Bank at Pittsfield, Illinois, informs local farmers of the latest prices offered for soybeans, shelled corn, and live hogs at the major Middlewestern markets. Photograph by John C. Hudson.

and Hewlett-Packard, have constructed research and development facilities here. A major national research facility, the Fermi Lab–National Accelerator Laboratory, is about 3 miles northwest of the East–West Tollway and Illinois Highway 59 interchange. Office parks abound in this area.

Twenty miles past the junction of I-290 and I-294, I-88 enters Kane County, which is the new office-park "frontier" of the Chicago metropolitan area. Continue west on I-88 past exits for *Aurora* (population 99,580). Although road signs still announce that I-88 is a research and development corridor, the transition to open land begins soon after the highway bends northwest onto higher ground just west of Aurora.

The broad, low rises in elevation are moraines produced by the Lake Michigan glacial lobe. But unlike the rough topographic texture of the forested, lake-border moraines crossed between Chicago and Naperville, these rolling uplands west of Aurora are smooth. Early Euro-American observers described them as prairies. The moraines closest to Lake Michigan were formed by ice advances that occurred less than 15,000 years ago, whereas the smoother moraines between Aurora and De Kalb represent ice margins as they stood approximately 19,000 years before the present.

Corn and soybean crops are the major land uses here today. Livestock, principally hogs and dairy cattle, once were important but they have been eliminated from the operations of farmers who now rely on cash-grain agriculture. A century ago, this area was part of the wheat belt of northern Illinois. Like most areas of Illinois and Iowa at this latitude, the country around De Kalb was settled by a mixture of New Yorkers and Pennsylvanians who brought wheat culture west with them in the 1840s. Corn Belt agriculture (raising corn to feed meat animals) was adopted by the 1880s after years of wheat-crop failure. Many other farmers turned to dairying at that time and this area was long an important supplier of milk to the Chicago market. Today, however, the old dairy barns and hog houses are either gone or rotting in disuse. Crops of corn and soybeans are trucked 50 miles south to a terminal on the Illinois River, poured into barges, and sent down the Mississippi to New Orleans for export overseas.

De Kalb (population 34,925) is separated from the Chicago metropolitan area by miles of cornfields, but its growing economy, based, in part, on the presence of Northern Illinois University (founded in 1895) illustrates the substantial growth in business in recent years in counties outside the metropolis.

Motels and restaurants are concentrated just south of the university campus on Lincoln Highway.

Day Two

THE MISSISSIPPI RIVER VALLEY
De Kalb, Illinois, to Ames, Iowa, 339 miles

For the next two days, the route will cross the Corn Belt of the Middle West. Many American and foreign tourists attempt to minimize the boredom they anticipate here by traveling as fast as possible along Interstate 80. But as a recent guidebook, *Traveling Interstate 80 with Otto* (1990), amply demonstrates, there are hundreds of fascinating things to see if only one bothers to get off the four-lane highway and into the landscape. I-80 is plagued by heavy truck traffic twenty-four hours a day. It is currently being rebuilt over much of the distance west of Chicago, which means long delays for the traveler. Besides these reasons for avoiding I-80, practically any other route is more interesting. But, for the sake of expediting the journey, we will take the interstate highway in a few places where the landscape can be appreciated at speed as well at a stop. One of those segments is the journey across Illinois.

De Kalb to Rock Island, Illinois, 126 miles, I-88

Rejoin I-88 at the Glidden Road entrance south of De Kalb and travel west. Ten miles west of De Kalb the highway crests a prominent rise that is part of the Bloomington end moraine formed in ice advances approximately 19,000 years ago. Traveling down the long slope toward *Rochelle* (population 8,770), I-88 intersects

De Kalb, Illinois, to Ames, Iowa

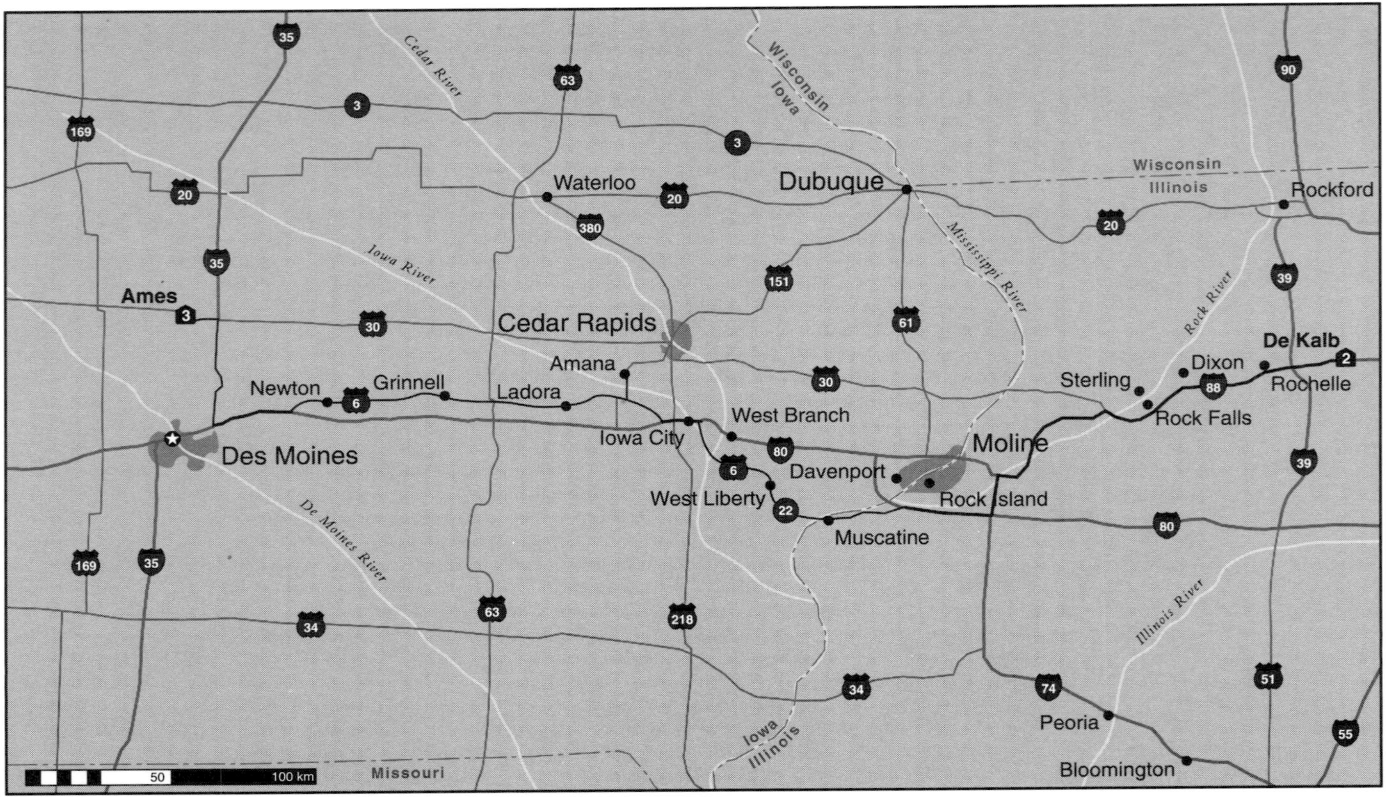

I-39, a newly completed highway linking cities north–south in Illinois. (Note: we stay on I-88 west.) To the south, near Bloomington (population 52,000) is a new Chrysler/Mitsubishi auto assembly plant; near Rockford (population 139,400) to the north, is an earlier Chrysler plant. These factories typify recent decentralization in the U.S. motor vehicle industry which has chosen locations in small and medium-sized cities outside major metropolitan areas.

Near *Dixon* (population 15,145) are former homes of two famous Illinoisians. The restored home of John Deere (b. 1804; d. 1886) is located in the hamlet of *Grand Detour,* 6 miles north of Dixon via State Highway 2. Deere was a young Vermonter who moved here in 1837 and began manufacturing iron plows. In 1847 he introduced the English-design steel plow and moved his growing business to Moline (population 43,200) on the Mississippi River, which is still the headquarters of Deere and Company. The birthplace of Ronald Reagan (b.1911), president of the United States, 1981 to 1989, is at *Tampico* (population 835), 13 miles south via Illinois Highway 88 at exit 41. Another of Reagan's boyhood homes is in Dixon. Routes to both are well-marked with brown information signs.

In this section I-88 is following the lowlands along the *Rock River.* The sandy bottomlands produce more abundant crops when irrigated, and this has led to the adoption of center-pivot irrigation systems. Dixon, *Rock Falls* (population 9,650), and *Sterling* (population 15,130), on the north bank of the Rock River, have manufacturing-based economies.

The Rock River, which rises in marshlands in central Wisconsin, is geologically a recent stream. The Ancient Mississippi River once flowed from northwest to southeast across this portion of western Illinois, cross-cutting what is now the Rock River, and eventually joined the channel of the present Illinois River in the central portion of the state. Ice sheets blocked this route approximately 21,000 years ago and the Mississippi then found the course it now occupies along the Illinois–Iowa border.

After entering Rock Island County (about 30 miles west of Rock Falls), a large meat-packing plant is visible in about 10 miles near the small town of *Joslin*. This is a beef and hide factory owned by IBP (about which you will see more at Dakota City, Nebraska). The cash-grain orientation of farmers in eastern Illinois is not shared by their counterparts in the western portion of the state. Livestock feeding was well established near here by the 1850s and the area has remained an important meat producer.

At the intersection of I-88 with I-80 east of Moline and Rock Island (84 miles west of De Kalb), follow I-80 east (compass south). I-80 crosses the Rock River on a high bridge and then crosses the *Hennepin Canal*. Built after 1900, this canal came into service at least fifty years too late and never had a significant economic role. It was built as a short-cut from the Illinois River at Hennepin to the Rock River about 3 miles west of the I-80 crossing. Its purpose was to bring traffic from the upper Mississippi River to Chicago without having to go south nearly to St. Louis to enter the Illinois River.

After traveling 6 miles south on I-80 (east), exit to I-280/I-74 west to follow the south bypass around Moline and Rock Island. I-280/I-74 starts paralleling the Rock River about 4 miles west of the I-80 interchange. The river is flowing at the base of some significant rock formations which are responsible for the higher elevation of the land surface on the south side of the valley. Highway exits pointing to *Coal Valley* (population 2,685) on the south side of the Rock River and *Carbon Cliff* (population 1,495) on the north may surprise travelers who do not expect to find coal in this area.

The latitude of Rock Island is roughly the northern limit of the lower Pennsylvanian coal measures (Spoon Formation) in the Middle West. While coal is far more abundant in the Carbondale Formation accessible farther to the south, coal mining was once a significant economic activity near Rock Island. These mines lie at the northern edge of the Eastern Interior Coal Field which is formed mainly around the Illinois Basin, a structure that reaches south to Kentucky. The Eastern Interior coals, which average 3 percent sulfur content by weight, far exceed the 0.7 percent federal air pollution standard that has been in effect for the past twenty years. Their future use remains in doubt.

SIDE TRIP TO VIEW THE MISSISSIPPI RIVER

After joining Illinois Highway 92 and crossing the Rock River, take the first exit north of the river (31st Avenue exit). Turn left at the stop sign and follow signs indicating *Sunset Park.* Go past the marina and follow the road toward the riverfront. Here is a very good water-level view of the Mississippi River. Restrooms and picnic facilities are available and so is an observation platform from which to view the main channel of the river that fronts the park. The grain elevator across the river is in Davenport, Iowa. To the north of the park are various river-oriented businesses.

Return via the same route to 31st Avenue and continue north through the city of Rock Island on Highway 92.

Travel 16 miles west on I-280/I-74, then exit at Illinois Highway 92 east (exit 11B). The highway crosses the Rock River and enters the city of *Rock Island* (population 40,550). Rock Island, neighboring *Moline* (population 43,200), and *East Moline* (population 20,150) constitute the Illinois side of the *Quad Cities* and account for about half of the conurbation's population. The Quad Cities have long been an important center for the manufacture of farm machinery. Like many other machinery-manufacturing areas of the United States, this one has suffered an extended decline in its employment base because of decentralization of the industry, corporate takeovers, and foreign competition. The abandoned farm machinery manufacturing plant of the J. I. Case Company along Highway 92 on the south edge of downtown Rock Island illustrates just one of the Quad Cities' economic woes.

Rock Island, Illinois, to Davenport, Iowa, 4 miles, Highway 67

Leave Highway 92 in downtown Rock Island at its junction with U.S. Highway 67 north. Highway 67 crosses the Mississippi River to Davenport via a toll bridge. This high bridge offers a good view of the river as well as the Quad Cities' riverfront areas. (It is impossible to stop here for taking photographs.)

Although the *Mississippi River* is the largest in North America, travelers who first see this legendary stream at Rock Island may be disappointed. Its volume here is less than one-twelfth that which it discharges into the Gulf of Mexico south of New Orleans. In fact, the Mississippi is smaller than the Tennessee River and less than one-fifth the maximum size of the Ohio River. The volume of water in the river's main channel north of St. Louis, Missouri, is insufficient even to permit barge navigation and, for this reason as well as flood control, the river was refashioned into a series of lakes behind dams. Twenty-seven dams and barge locks punctuate the river between St. Louis and St. Paul, Minnesota. Lock and Dam No. 15 is in downtown Davenport/Rock Island; No. 16 lies 40 miles south at Muscatine (population 22,400) while No. 14 is only a few miles upstream near the city of Bettendorf (population 28,100). The locks and dams are under the operation and control of the U.S. Army Corps of Engineers. Water transportation is one of the most massively subsidized businesses in this country. Nearly all waterways have been built and are operated at taxpayers' expense, but are used practically free of charge by a few companies in the barge business.

This is an unusual stretch of the Mississippi because the river is flowing more to the west than to the south at this point. Its channel bends to the west here, before finding a southern outlet near Muscatine, because this is where it encounters the edge of the resistant Devonian and Pennsylvanian limestones (the latter contains the coal measures exposed along the Rock River just to the east). Reservoirs behind the dams around Davenport and Rock Island drown a rocky stretch of rapids in the river channel which

The Mississippi River north of St. Louis consists of a series of barge locks and dams which control water level in the river and also serve as a flood control system during periods of high water. Photograph by John C. Hudson.

posed low-water navigation problems for nineteenth-century steamboats.

Both Rock Island and *Davenport* (population 95,335) are authentic river towns of the mid-nineteenth-century golden age of steamboating on the Mississippi. Their tall, brick storefronts, narrow streets, and stately old homes immediately suggest a historic past more like that of longer-settled Ohio or Indiana rather than Illinois or Iowa. Indeed, the river towns are about a generation older than their inland counterparts. More than two dozen towns were founded in this stretch of the river between 1826 and 1842, of which more than half failed. Rock Island (founded in 1836) and Davenport (1832) became natural rivals for prominence at these rapids of the Mississippi.

The river trade declined after railroads began reaching west of Chicago in the 1850s. The first railroad bridge anywhere in the West was constructed across the Mississippi between Rock Island and Davenport between 1854 and 1855. Within eighteen months two-thirds of the traffic these towns had sent downriver to St. Louis was moving west to Chicago via the Rock Island Railroad.

Davenport and adjacent *Bettendorf* account for most of the population on the Iowa side of the river. Farm machinery manufacture, heavy-equipment industries, and meat packing have been the traditional sources of employment here, although, as on the Illinois side, the economy has shifted away from manufacturing in recent years.

Davenport to Muscatine, Iowa, 31 miles, Highway 22

After leaving the Highway 67 toll bridge across the Mississippi, follow U.S. Highway 61/Iowa Highway 22 south along the industrial south side of Davenport, past the large and abandoned meat-packing plant along West River Drive. At the divergence of Highways 61 and 22, follow Highway 22 toward Muscatine.

Although railroads put the Mississippi River steamboats out of business in the nineteenth century, the river has recently gotten its revenge. The export grain business of the United States boomed in the 1970s due to favorable agricultural policies and good foreign markets. Corn and soybeans that once moved by rail to the east began to be diverted down the Mississippi River for export. By the 1980s this caused the abandonment of many local railroad lines, which had already lost all of their non-grain traffic to trucks.

Two barge-loading grain terminals ("river elevators") are located 2 miles south of the Highway 22/I-280 interchange. Most of the corn and soybeans arrives here by truck from eastern Iowa or western Illinois, but grain arriving by rail from longer distances also is dumped here and loaded aboard barges. Oil and coal coming upriver and molasses being sent downriver also are handled here. South toward the sleepy river town of *Buffalo* (population,

1260) are two large Portland cement plants which are based on rock quarried in the pits west of the highway. All of these industries, as well as the coal-fired electric-power–generating station south of Buffalo, rely on the river to float heavy bulk commodities in or out.

After entering the city limits of *Muscatine* (population 22,880) make the first right turn and go up the hill to Weed Park (no signs mark this approach). A reasonable view across the Mississippi River floodplain can be had from any of the observation platforms in this pleasant park overlooking the valley. The Mississippi flows in a rather narrow valley here, atypical of most of the river's course. In fact, the river has occupied this location for a mere 21,000 years. The Ancient Mississippi River flowed across northeastern Illinois and occupied the present-day south-flowing portion of the Illinois River, but that route became blocked by glacial ice north of Rock Island. The segment seen here was one of the last additions forming the river which separates Illinois from the Iowa and Missouri we know today.

Return to Highway 22 and then take Business Highway 22 to see downtown Muscatine. Clamshells dredged from the river once were used to make shirt and dress buttons, a local industry which gave Muscatine the nickname, "Pearl Button Capital" of the United States. The Muscatine melon, a tasty muskmelon (cantaloupe) grown on the sandy bottomland soils near the city, also brought recognition. Today Muscatine is a diversified manufacturing city, but its historic flavor remains. The Muscatine County courthouse, the old jail on one corner of the courthouse square, and other buildings in the city's downtown give an authentic impression of a Mississippi River town that has changed slowly over the years.

Continue through Muscatine on Business Highway 22 and rejoin Highway 22 at the western edge of the city.

Muscatine to Amana, Iowa, 60 miles, Highways 22, 70, and 6

After rejoining Highway 22, travel west 10 miles to the intersection of Iowa Highways 22 and 70. In this section of Highway 22 the road travels through upland farm fields and then, about 10 miles west of Muscatine, descends to cross the *Cedar River* on a low bridge. The thick forest on both sides of the highway is growing on a former lake bottom, that of glacial *Lake Calvin.* The lake was formed when ice sheets blocked the southward flow of the Iowa and Cedar rivers. It drained when the ice retreated, leaving a low, swampy land surface that extends for several miles to the west. Soon, however, the road starts climbing through surrounding corn land once again.

Continue straight ahead (north) at the intersection of Highways 22 and 70 and follow Highway 70 into *West Liberty* (population 2,935). This former railroad junction is now a meat-packing town. The meat packed here is turkey, and those wishing to see a lunchmeat factory can reach it by following the trail of white feathers along streets leading to the plant, which is on the west side of town. West Liberty's main business street (3rd Street) retains the appearance of the Corn Belt trade-center town of old.

Take U.S. Highway 6 west at its junction with Highway 70 in West Liberty, staying alert for signs indicating County Highway X40 which heads north from Highway 6 on the north edge of West Liberty. Follow Highway X40 north for 6 miles. Just past the overpass over I-80, turn left off X40 onto Highway F44, which has been officially designated the Herbert Hoover Highway (HHH).

Follow the HHH west 5 miles into the village of *West Branch* (population 1,910), the birthplace of Herbert Clark Hoover (b. 1874; d. 1964), president of the United States, 1929 to 1933. Turn left (south) in downtown West Branch onto Cedar County Highway X30 to see the Hoover Presidential Library and Museum. A school, Quaker meeting house, and other buildings dating from Hoover's early years in West Branch are located on the grounds.

Continue south on Highway X30 to its junction with I-80 west and enter the interstate highway.

Highway I-80 has a rolling profile here, as do all roads crossing southeastern Iowa. Glaciers covered parts of Iowa several times during the Pleistocene. In the intervals between glaciations, westerly winds continually eroded surface sediments only to have gravity redeposit them once more. Thick layers of these fine-grained, windblown materials, known as loess, blanket much of Iowa. Where the land surface is covered with loess the topography is more rolling, typically creased with many short, steeply sloping valleys. An average of 16 feet of loess covers this portion of the state.

Ten miles after entering I-80, the highway passes along the north edge of *Iowa City* (population 59,740). To reach Iowa City, home of the University of Iowa (founded in 1847), leave I-80 at exit 244 (Dubuque Street) and travel south into the city. Iowa City was the first capital (1846 to 1857) of the state. The former capitol building, an interesting Greek Revival structure constructed largely of limestone quarried from the banks of the Iowa River and featuring a gold dome, is on Dubuque Street in the center of town surrounded by the university's campus.

Just past exit 244, I-80 descends a steep valley to cross the *Iowa River.* At exit 240 in *Coralville* (population 10,350), leave I-80 and follow signs for U.S. Highway 6 west. Follow Highway 6 through 12 miles of rolling, loess-mantled hills to the intersection with U.S. Highway 151 at *Homestead.* This is the beginning of the Amana Colonies. (Bill Zuber's Restaurant in Homestead offers excellent Amana-style food.)

Follow Highway 151 for 3 miles north, across the Iowa River, to *Amana* (population 600), the largest and oldest of the seven Amana villages. The villages and surrounding farmlands along the Iowa River were developed by the Amana Society after 1857 when the group moved here from their former colony near Buffalo, New York. (The group has no affiliation with the Amish, an Anabaptist group, despite the similarity in name.) Ownership of land and buildings was in the name of the community. The large size of many of the attractive, old brick and sandstone buildings in the villages reflects the former practice of dormitory housing which deemphasized the role of the family. Social and economic functions were separated in the 1930s, after which time the villages became places of residence for families who continued to work on the large Society-owned farms. Woolen mills, furniture factories, wineries, and a variety of other small manufactories were located in the several villages. Development of the Amana Refrigeration Company (now owned by the Raytheon Corporation) marked another successful economic venture for the Society.

The role of tourism has grown steadily since completion of I-80 in the 1960s. But Amana has managed to retain its charm despite the steady parade of automobiles and tour buses heading into the villages. Hearty, Amana-style meals (resembling Pennsylvania German cuisine) are served at the Ronneburg Restaurant and Ox Yoke Inn and at half a dozen newer eating establishments. The old winery used to specialize in local favorites such as dandelion or rhubarb wine, but today's many Amana wine cellars follow the model of New York or California's wine-grape regions. The main farm at Amana has been converted into a convention center. Craft shops abound, of course.

Amana to Ames, Iowa, 118 miles, Highways 220 and 6

Travel west on Iowa Highway 220 through *Middle Amana, High Amana,* and *West Amana.* These villages are less often visited and they have changed less in appearance. Return to U.S. Highway 6 at *South Amana* and continue west on Highway 6. The highway skirts *Marengo* (population 2,300), seat of Iowa County and local trade center. Six miles west of Marengo is *Ladora* (population 250). The imposing facade of the Classical Revival-style Ladora Savings Bank, long abandoned, in a town so small, suggests people's former aspirations. In addition to hard times on the farm, these towns along Highway 6 suffered from the disappearance of tourism after I-80 was constructed. Highway-oriented businesses have moved to the freeway interchanges. A close look will reveal grass growing in the

Former station of the Minneapolis & St. Louis and the Rock Island railroads at Grinnell, Iowa. The station is abandoned and the tracks are seldom used these days. The state of Iowa has about half as many miles of railroad today as it had in 1970. Photograph by John C. Hudson.

pavement of Highway 6 as it stretches its lonely way west from here.

Twenty-eight miles west of Ladora is *Grinnell* (population 8,900), founded in 1854 by Josiah Bushnell Grinnell, an abolitionist clergyman from Washington, D.C. Grinnell College was organized by the Congregational church at Davenport in 1847 but was moved to Grinnell in 1859. The town's Yankee heritage is typical of middle Iowa and also of the northern portion of the state, but south of here settlements were established by natives of Indiana or Ohio and the New England influence is proportionally less important. Grinnell is also the site of Louis Sullivan's famous design for the Merchants National Bank, completed in 1914.

West of Grinnell the highway crosses the *North Skunk River,* another of the roughly parallel, southeast-flowing streams that are

SIDE TRIP TO DES MOINES

Turn south on I-235 at its interchange with I-80 (exit 138 from I-80) on the northeast corner of the city of Des Moines. The city's handful of tall buildings can be seen immediately as I-235 drops downgrade toward the city. Des Moines began as a military post, Fort Des Moines, in 1847. It became the state capital in 1857 when the seat of government was removed here from Iowa City. Des Moines is a financial and insurance center as well as the principal trading point and medical service center for many Iowans. The impressive state capitol building and the Iowa Historical Society museum are on the east bank of the Des Moines River near the city's center.

In addition to a fair collection of the usual urban parks, Des Moines has a most unusual, working exhibition of Middlewestern farming. The 600-acre *Living History Farms* is on the northwest side of the city, and it is open from May through October. (For information, telephone 515-278-5286.) Follow I-235 west to its junction with I-35, then take I-35 north to exit 125. The farms show agriculture as it was practiced by the Indians and by farmers in the Euro-American pioneer phase. History is then updated in an early 1900s version of an Iowa farm which is complemented by a display of modern agriculture. Various historic buildings typical of town settings also are included.

cut into the loess-mantled surface of eastern Iowa. Nineteen miles west of Grinnell is *Newton* (population 14,790), the seat of Jasper County. Newton is a fairly prosperous industrial city, due in part to the employment base provided by the Maytag washing machine company. At one time, five washing machine manufacturers operated here. The Jasper County courthouse, set in a central square in downtown Newton, is a classic piece of Middlewestern town design.

West of Newton the traveler has no alternative to I-80. Go to the Iowa Highway 14/I-80 interchange on the south edge of Newton and travel west on I-80 toward Des Moines. About 5 miles west of Newton, I-80 follows the valley of the *South Skunk River* for 10 miles. The highway leaves this valley near *Mitchellville* (population 1,670) and climbs steadily to a higher elevation. The crest of the grade is reached at the road overpass at exit 149.

The transition to high ground at this point marks the southeastern edge of the Des Moines lobe of the Late Wisconsin ice sheet. The hill is the Bemis moraine which was deposited at the edge of and underneath the ice sheet 14,000 years ago. The Des Moines lobe is a more recent feature than the loess deposits that cover eastern Iowa, and hence glacial materials cover the loess here. A change in the landscape's appearance is obvious west of Mitchellville. The land surface is gently rolling, large grain elevators appear on the horizon, and the proportion of the land used for cash crops (principally corn and soybeans) increases.

Take I-35 north at its intersection with I-80 in northeast Des Moines. Away from the edge of the Bemis moraine, the surface of the Des Moines lobe is comparatively flat. Soybeans are the most important crop here in most years. Continue north on I-35 about 25 miles to *Ames* (population 47,200), home of Iowa State University (founded in 1858). Motels and restaurants are available just west of exit 111 along U.S. Highway 30 west toward the city of Ames. The franchise-food strip, with more motels, follows U.S. Highway 69 (north of its intersection with Highway 30) toward the central business district of Ames.

Ames, Iowa, to Norfolk, Nebraska

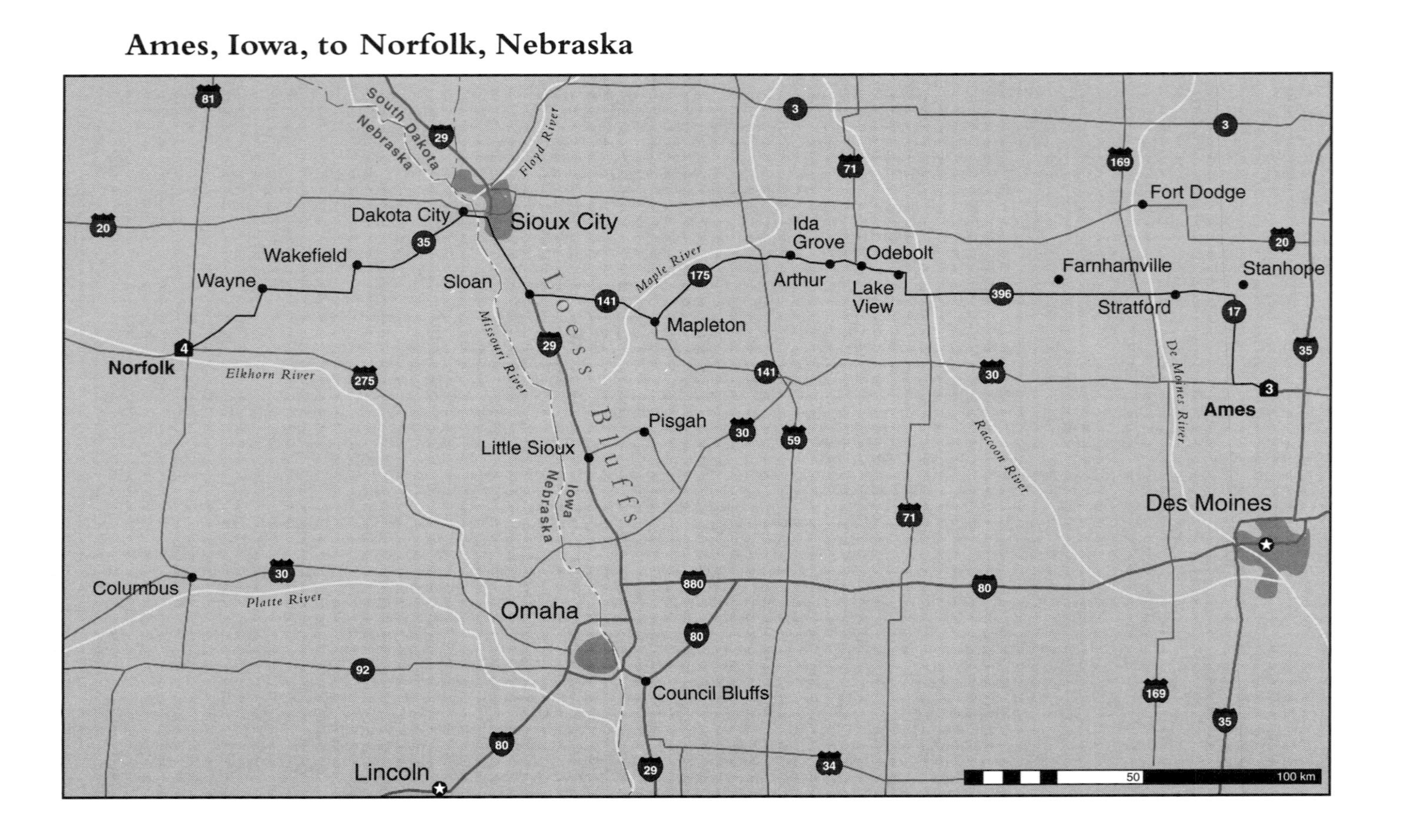

Day Three

THE MISSOURI RIVER VALLEY
Ames, Iowa, to Norfolk, Nebraska, 265 miles

Most of this day will be spent traveling the backroads of western Iowa and eastern Nebraska. The route leads across the glaciated plains of the Des Moines lobe, crosses the thick loess deposits of western Iowa, descends briefly to the Missouri River lowland for a look at Sioux City, and then reenters rolling country side in the loess hills of eastern Nebraska. This far west it is still a Corn Belt landscape, but a more recently settled one that has never attained a large population. It is a land of small towns, large farms, and long vistas.

Travel here is different from that along the interstate highways. Those who assume that McDonalds, Burger King, Taco Bell, and Pizza Hut are ubiquitous in the United States will discover otherwise today. Anyone who cannot live without this sort of food is advised to stock up before leaving Ames or wait until Sioux City for your next meal. There is no danger of going hungry, however, for those willing and eager to sample the meals offered in small-town cafes. Every town has at least one gas station, although many are closed on Sundays.

Ames to Stanhope, Iowa, 28 miles, Highways 30 and 17

Travel west from Ames on U.S. Highway 30. Along the highway are miles of fields where crop experiments are being conducted by agricultural scientists at Iowa State University. The bumper crops of corn harvested around the world today owe much to plant-breeding expertise at universities such as Iowa State. Hybrid corn—which for all practical purposes now includes all commercially produced corn—was first marketed in central Iowa in the 1930s.

Corn produces both male and female elements on the same plant. Male cells are produced in the tassels at the top of the plant, while the female cells are found in the ear shoots that develop along the stalk. The production of hybrid seed corn involves planting adjacent strips of genetically dissimilar varieties and then removing the tassels from the plants intended to produce the ears of corn. Continued experimentation has resulted in corn yields today that more than double what they were in the 1930s. The stalks of hybrid corn plants are stronger than the old, open-pollinated varieties and thus are more easily picked by machine.

Turn right (north) onto Iowa Highway 17 at its intersection with Highway 30 about 10 miles west of Ames. A hybrid seed-corn plant is on the left side of the highway just north of the intersection. Locally grown seed is cleaned, dried, sorted, and packaged for sale in many small factories such as this one. Just past the seed factory is a large grain elevator owned by the West Central Co-operative at Ralston, Iowa. Farmer-owned co-ops such as this one now deal millions of bushels of corn and soybeans in the international market. This elevator stores more than three million bushels of grain and can load a seventy-five-car train of 100-ton grain cars without having to replenish its stocks.

Highway 17 is here passing over a succession of low, parallel, east–west-trending ridges, known as a washboard moraine. Each ridge represents a brief readvance of the glacier that created the Des Moines lobe. Highway 17 crosses the more prominent crest of the Altamont moraine roughly where Boone County Highway E18

Corn products are a major industry in Iowa. This plant at Cedar Rapids manufactures corn sweeteners (which are now widely used in place of traditional sugar products in prepared foods) and produces alcohol for blending into "gasohol," an automobile fuel. Photograph by John C. Hudson.

intersects. The Altamont was deposited within a few hundred years after the glacier's maximum advance created the Bemis moraine. The entire ice sheet wasted northward to the vicinity of Algona (population 6,000) near the northern edge of Iowa by 13,000 years before the present. The glacier was retreating northward an average of 750 feet per year at that stage, near the end of the episodes of continental glaciation.

Stanhope to Mapleton, Iowa, 115 miles, Highway 175

Turn left (west) onto Iowa Highway 175 at its intersection with Highway 17 a mile south of Stanhope. The land surface here is a swell-and-swale topography left behind by the glacier. Natural drainage is inadequate much of the year and this requires digging ditches and laying underground tile to remove excess water. Ditches are visible along field margins, sometimes next to the highway. These rolling fields of the Des Moines lobe are among the highest-valued farmlands in Iowa. Corn and soybeans bring a greater dollar return per acre than anything else the land here can produce.

Soybean fields east of Farnhamville, Iowa, on the Des Moines lobe. These are among the more productive acres in the United States, although the land was once a wet prairie and had to be drained before agriculture could be made productive. Photograph by John C. Hudson.

All of the soybeans and most of the corn are sold to local elevators. Cash-grain farming has replaced livestock feeding, a fact reflected in the abandoned barns still standing alongside new grain-storage bins on farms along the highway.

Stratford (population 715) is a well-preserved small trade center. Note, however, the telemarketing firm along the highway in this town, a sign of the new types of businesses located in seemingly remote and inaccessible places. West of Stratford, Highway 175 makes two 90-degree bends within a mile as it follows the section lines of the survey grid. Most of the road system in the rural Middle West originated as "section-line" roads which were gradually upgraded to highways.

Four miles west of Stratford the highway descends a steep hill down to the bridge over the *Des Moines River*. This river, which originates in southern Minnesota, runs along the axis of the Des Moines lobe. It has a deep valley over most of its course across north-central Iowa. At the west side of the bridge a gravel road (marked for Carlson Recreation Area) turns off to the left to follow along the river for several miles. A thick forest surrounds the road and is typical of the floodplain forest in this region of upland prairies. Trees along the river were cut for firewood and building materials in the early settlement years, although today the forest cover has more than reestablished itself.

West of the Des Moines River is *Dayton* (population 820), a town nestled in the hills. Drainage ditches reappear once the road returns to the uplands. The highway then skirts *Harcourt* (population 300), *Gowrie* (population 1,030), and *Farnhamville* (population 400). These towns along the Chicago & Northwestern Railroad were created by townsite promoters who selected quarter-section (one-fourth square mile) tracts of land 7 or 8 miles apart. The highway bypasses the towns not because they are congested and need a bypass, but rather because the highway is following a straight-line course along the survey grid. Town centers are on the railroad, typically either a quarter- or three-quarters of a mile away, in the center of the parcel of land the townsite promoter acquired. Iowa's grid of roads and its network of small towns, both highly regular, are slightly out of kilter for this reason. Major highways often intersect in cornfields instead of in towns.

Turn right (north) on Calhoun County Highway P21 to get a closer look at the massive grain elevators at Farnhamville. They belong to the local Farmers Cooperative and they can store more than nine million bushels of grain. Corn and soybeans are shipped by the trainload, either to the West Coast for export to the Far East, to corn- or soybean-processing plants in the Middle West, or sometimes to the Mississippi River for barge loading. This particular farmers' co-op also happens to own another eight grain-storage terminals in the area and sells fertilizer and agricultural chemicals to local farmers. Note the electronic sign in front of the co-op's headquarters, offering up-to-the-minute quotations on corn and soybean futures trading in Chicago.

Return to Highway 175 and continue west through *Lohrville* (population 450) and *Lake City* (population 1,840). The latter has a tree-shaded central square of the sort that has long been home to Saturday-evening band concerts and ice cream socials in the small towns of the Corn Belt. Continue west past *Auburn* (population 300) whose welcoming sign, "Just the Other Side of Everywhere," no doubt summarizes the feeling of remoteness in many small towns today.

As the highway passes *Lake View* (population 1,300) the land surface changes once again. This is the western edge of the Des Moines lobe. Beyond the crossing of the *Boyer River,* 3 miles west of Lake View, the glacial deposits are gone and the land surface is once again formed on a loess cover, here approximately thirty feet thick. Livestock farms appear as soon as the transition to hilly land is made. This is the beginning of an almost continuous zone of cattle-feeding specialization that extends west to the Nebraska Sand Hills.

Small towns can surprise you. Highway 175 follows the rolling hills past *Odebolt* (population 1,200), through an area that once specialized in popcorn production. The Parthenon-like Odebolt State Bank, just off the highway, is worth seeing, for one reason because it still is a bank. Unfortunately, the "Cracker Jack" logo has been erased from Odebolt's grain elevator. *Ida Grove* (popula-

SIDE TRIP TO THE LOESS HILLS SCENIC BYWAY

A closer look at the Missouri Valley loessial bluffs can be had by traveling 32 miles south on I-29 to its junction with Iowa Highway 301. Then take Highway 301 east to reach the village of *Little Sioux* (population 250). Continue east from the village via a secondary road to reach the Scenic Byway, which follows the foot of the bluffs. Harrison County road F20 crosses the Scenic Byway and then climbs the bluffs. At the top of the grade is a parking lot for the Murray Hill Scenic Overlook; a trail leads from there along the crest. Here is an excellent view of the steep bluff face as well as the floodplain of the Missouri River. The Little Sioux River is channelized here, as are most of the streams entering the Missouri in this section. The scenic overlook is also accessible from Iowa Highway 183 by turning west on Harrison County road F20 at Pisgah.

tion 2,400), 12 miles to the west, was a normal-looking Iowa town until it was given a Camelot theme-park atmosphere by a local industrialist who happens to like moats and castles.

Continue west on Highway 175 as it parallels the Maple River through *Battle Creek* (population 815) and *Danbury*. The thickness of the loess deposits increases toward the west. The Missouri River's floodplain was the source of this fine-grained material that can stand, exposed, as an almost vertical wall. The formations get more spectacular as the Missouri's floodplain is approached. Cuts visible along the roads that lead down into the Maple Valley removed at least fifty feet of loess.

Mapleton to Sloan, Iowa, 25 miles, Highway 141

Follow Highway 141 west at its intersection with Highway 175 just north of Mapleton. Highway 141 immediately climbs the steep hills bordering the Maple River. Approximately 8 miles beyond Mapleton the highway descends to the floodplain of the *Little Sioux River* and passes through *Smithland* (population 250). Four miles west of Smithland the highway reaches the bluffs at the edge of the Missouri River floodplain. (Note: A pull-off area to the left of the highway allows parking at the crest of the bluffs. From here

Loess bluffs along the edge of the Missouri River floodplain east of Little Sioux, Iowa. The steep bluffs were deposited from winds which eroded fine-grained sediments from the Missouri's flat, bordering plain. The bluff line is nearly continuous along the river in Missouri and Iowa. Photograph by John C. Hudson.

can be seen the line of steep bluff faces to the north and south. The view west is toward Nebraska.) The bluffs are a product of wind action on exposed sediments of the Missouri River floodplain just beyond. The line of hills rises dramatically from the edge of the floodplain.

Continue west on Highway 141 through *Hornick* (population 250). The Missouri River bottomland is flat, fertile, and mostly used to raise crops of corn and hay. At *Sloan* (population 940) Highway 141 ends at the junction with I-29.

Sloan to Sioux City, Iowa, 21 miles, I-29

The bluffs on both the Iowa and Nebraska sides of the Missouri River are visible from I-29 as the valley narrows in width to the north. The smokestacks to the west of I-29 near *Salix* (population 400), 7 miles north of Sloan, identify two coal-fired electricity-generating stations belonging to the Iowa Public Service Commission. They are located in the *Port Neal* industrial area. Port Neal's complex of industries reflects the region's agribusiness economy. Included are a soybean-processing plant, an ammonia terminal, a large fertilizer manufacturer (which receives trainloads of potash from Saskatchewan in Canada), and a gelatine plant and a tannery which use byproducts of local meat-packing industries.

Leave I-29 at exit 143 and take U.S. Highway 75 north. The Big Sioux Grain Terminal west of this exit receives grain for barge shipment south on the Missouri River. *Sioux City* (population 80,500), perhaps the most inland "harbor" in the United States, is at the head of barge navigation on the Missouri. Rail shipment of grain (as well as rail receipt of low-sulfur coal from Wyoming) greatly overshadows the river's importance today, however. Sioux City was an important river town in the era of Missouri River steamboating. Trading posts were built here in the 1850s. But Sioux City's rise to importance came late in the nineteenth century when major meat-packing firms located here.

Follow Highway 75 north to the *Floyd Monument,* an obelisk commemorating the death here in 1804 of Sergeant Charles Floyd, Jr., a member of the famous two-and-a-half-year Lewis and Clark expedition across the continent. The view from the monument is north toward downtown Sioux City and west across the Nebraska side of the floodplain.

Continue north on U.S. 75 to Leech Avenue in Sioux City. Turn west on Leech, traveling past the sprawling, abandoned hulk of Sioux City's largest packing plant. The Midland Packing Company (later, Swift & Company) constructed the complex in 1919. An attempt has been made to convert the building into an indoor amusement park and shopping center.

Cunningham Street intersects Leech Avenue a block west of the abandoned packing plant. Turn left (south) on Cunningham to see the heart of Sioux City's old stockyards district. The Sioux City Livestock Exchange conducts a feeder pig auction here every Tuesday. Farther south on Cunningham can be seen the active meat-packing plant of the John Morrell Company. Cunningham Street's remaining saloons (such as the Stockyards Plaza) suggest the once-robust life along this street.

The large brick building at the bend in Cunningham Street is the *Livestock Exchange* building. Its tile floors and wood-paneled corridors see much less traffic than in former times, but within the building can still be found an authentic stockyards restaurant (Silver Steer Restaurant and Bar), a branch of the Sioux City post office, and a comfortable "Ladies Lounge" for the wives and daughters of ranchers and stock buyers who have come to Sioux City on business.

Returning on Leech Avenue to U.S. Highway 75 (Lewis Boulevard), follow signs leading to U.S. Highway 20B west. Highway 20B crosses above the stockyards and railyards on a viaduct into the Sioux City central business district. Turn north on Nebraska Street to reach the heart of the city.

On Douglas Avenue between Sixth and Seventh streets are two interesting examples of Middlewestern urban architecture. City Hall, at Sixth and Douglas, exemplifies the Richardsonian style in vogue during the 1890s. Many Middlewestern cities on the rise at that time incorporated buildings that followed the Romanesque designs of Henry Hobson Richardson (b. 1838; d. 1886). On the northwest corner of the same block is the Prairie School Woodbury County Courthouse constructed between 1915 and 1918, the work of William Steele, who brought Louis Sullivan's architectural innovations to Sioux City. A spectacular dome of stained glass dominates the main-floor rotunda. Both buildings reflect Sioux City's turn-of-the-century aspirations to be a major metropolis.

Take Pearl Street south to its intersection with the highway and follow signs to U.S. Highway 77 south; take Business Highway 77/Business Highway 20B to South Sioux City, Nebraska, across the *Missouri River.*

The "Big Muddy," as the Missouri River is sometimes called, drains one-sixth of the United States. It has the largest drainage basin of all the rivers tributary to the Mississippi, and includes those of the Platte and Yellowstone, both major rivers, as well. But the Missouri flows through a semiarid region that typically produces less than two inches of runoff per year and consequently its volume of flow is not particularly large among U.S. rivers. At Sioux City, the Missouri carries about as much water as the Ohio River does at Pittsburgh, Pennsylvania, but the Missouri's upstream drainage area is more than twenty times larger.

Sioux City marks a significant change in the nature of the Missouri River. Upstream (west) the river has been impounded by the U.S. Army Corps of Engineers behind six giant, earth-fill dams that have flooded 100-mile-long reservoirs as far west as central Montana. Created for flood control, power generation, and irrigation, the great reservoirs have been only marginally successful other than for flood control; they have, however, created water-oriented recreation areas in the Plains. Downstream (south) from Sioux City there are no dams or reservoirs. Instead, the Missouri has been rebuilt into a navigation channel for barge traffic, its banks and floodplain lined with artificial levees. The Big Muddy thus bears little resemblance to its former lazy, unpredictable self and is now primarily a human-engineered system for moving water across country.

Sioux City, Iowa, to Norfolk, Nebraska, 79 miles, Highways 20B and 35

Business Highway 20B follows First Avenue in *South Sioux City* (population 9,670), an old-fashioned commercial strip common in western towns. The large elevators south of the city are part of the Con-Agra oat mill, another locally important agricultural processing industry. Just north of the limits of *Dakota City* (population 1,470) is the enormous meat-packing plant and headquarters offices of Iowa Beef Processors (IBP). The declining stockyards and abandoned packing plants of Sioux City constitute a sort of museum of the Middlewestern packing town that thrived until the 1960s, whereas IBP's sterile-looking complex of buildings (unrecognizable as a slaughterhouse) north of Dakota City represents the modern industry standard.

IBP does a $10 billion annual business and accounts for about one-fourth of the beef production of the United States; the company is expanding into pork operations as well, and currently holds a 12 percent share of the nation's pork market. Founded in 1960, IBP's Dakota City headquarters now oversees the operations of eleven beef plants and six pork plants concentrated in Iowa, Kansas, and Nebraska, but spread from Illinois to Texas to Washington State. Each of the plants prepares meat in boxed form for sale to grocery and hotel chains and to other meat processors. Bone, fat, trimmings, hides, and proteins are recovered for remanufacture, as in the past. The sight of 100 refrigerated trucks, engines running and waiting to load at the Dakota City plant, suggests the scale of the meat industry today.

Continue south past the IBP complex on U.S. Highway 77–Nebraska Highway 35. Take Nebraska Highway 35 west from Dakota City. Highway 35 steadily climbs across the Missouri River lowland. A row of livestock-feeding farms is visible at the western edge of the flat valley portion. Approximately 7.5 miles west of Dakota City, Highway 35 enters the loess hills of eastern Nebraska. Thirteen miles from Dakota City the road is still climbing the loessial and glacial-till surface, which here accumulated to a thickness of 200 to 300 feet, atop the Cretaceous shales. Fifteen miles from Dakota City, Highway 35 swings west into typical eastern Nebraska loess-hill countryside, with a predominance of stock-feeding farms. Corn and hay are the major crops.

The unmarked buildings at the crest of the hill just east of *Wakefield* (population 1,080) belong to an egg farm. "Any time, any place" is the slogan of the Milton G. Waldbaum Egg Company, which delivers eggs nationwide from Wakefield. Turn right from Highway 35 onto First Street in Wakefield. The street once was a railroad track. The former railroad depot and grain elevators retain their original locations, only now they line a city street. Main Street in Wakefield crossed the railroad tracks near the depot, a typical town layout in the western plains. Wakefield's brick streets lined with small storebuildings also are typical of the moderately successful trade-center town of seventy-five years ago. Note the community bulletin board in the middle of the town's major street intersection.

Return to Nebraska Highway 35 and continue through rolling loess hills via *Wayne* (population 5,140) to Norfolk. The industrial buildings north of Highway 35 on the approach to Norfolk belong to the Nucor steel mill, one of the modern mini-mills that is changing the nature of the steel industry. Raw materials are received in the form of scrap metal. Several "ferrous recyclers," a modern euphemism for junk yards, in and around Norfolk channel raw materials to the mill. Steel produced at Norfolk is trucked to numerous metal-fabricating plants, principally those that produce agricultural implements and livestock-feeding equipment within a several-hundred-mile radius.

Norfolk (population 21,475) is northeastern Nebraska's largest city. Its main street (Norfolk Avenue) acquired a mall-like appearance when trees were planted in the median strip, an ahistoric "beautification" that gives the street a more Eastern look. In addition to its new role as steel town of the Plains, Norfolk remains a major trade center for the area.

Motels and restaurants are clustered near the intersection of U.S. Highways 81 and 275 at the southwest corner of Norfolk. (The Grainary Restaurant north of this intersection offers good food in a setting evocative of the early twentieth-century Middle-western farmhouse.)

Norfolk (elevation 1,527 feet, as opposed to 1,117 feet in Sioux City and 921 feet in Ames) is the end of Day 3 of the journey and it also marks a significant transition in the landscape. This city lies near the western limit of continental glaciation. West of here the land surface is one of long, gentle slopes that lack the rough texture just seen in northeastern Nebraska. Norfolk is situated slightly east of the 98th meridian, which is sometimes taken as a convenient eastern boundary of the Plains regions. It also lies close to the border between the moist and dry climates. To the west, potential evapotranspiration exceeds precipitation, which means that, based on climatic factors alone, growing plants will produce more vegetative matter if the fields are irrigated. This city thus marks passage from the traditional Corn Belt of the Middle West to the more arid, sparsely settled rangelands of the Great Plains.

Day Four

THE SAND HILLS
Norfolk to Ogallala, Nebraska, 334 miles

Most of this day will be spent in the *Nebraska Sand Hills,* a 20,000-square-mile (52,000-square-kilometer) grassland on dune sand in north-central Nebraska. The area is sparsely populated. The largest town between Norfolk and Ogallala is Burwell (population 1,280); no towns with as many as 1,000 inhabitants are to be found west of there. Gasoline stations and cafes are available in most communities, but the level of service is rudimentary. Tourists tend to overlook this region, but it is great country to explore.

Norfolk to Dunning, Nebraska, 170 miles, Highways 275, 70, and 91

Take U.S. Highway 275 west from Norfolk. The highway follows the gently sloping valley of the *Elkhorn River.* Towns here were created by the Fremont, Elkhorn & Missouri Valley Railroad, which opened the area to agricultural settlement in the early 1880s. Dryland cornfields now alternate with center-pivot irrigation of corn or alfalfa in this zone of moderate drought risk. At *Tilden* (population 895), 21 miles west of Norfolk, a large farmers' cooperative exchange is spread across the town and features irrigation equipment, grain storage, chemical fertilizers, and feed-mixing facilities which illustrate the mixed nature of agriculture in this

Norfolk, Nebraska, to Ogallala, Nebraska

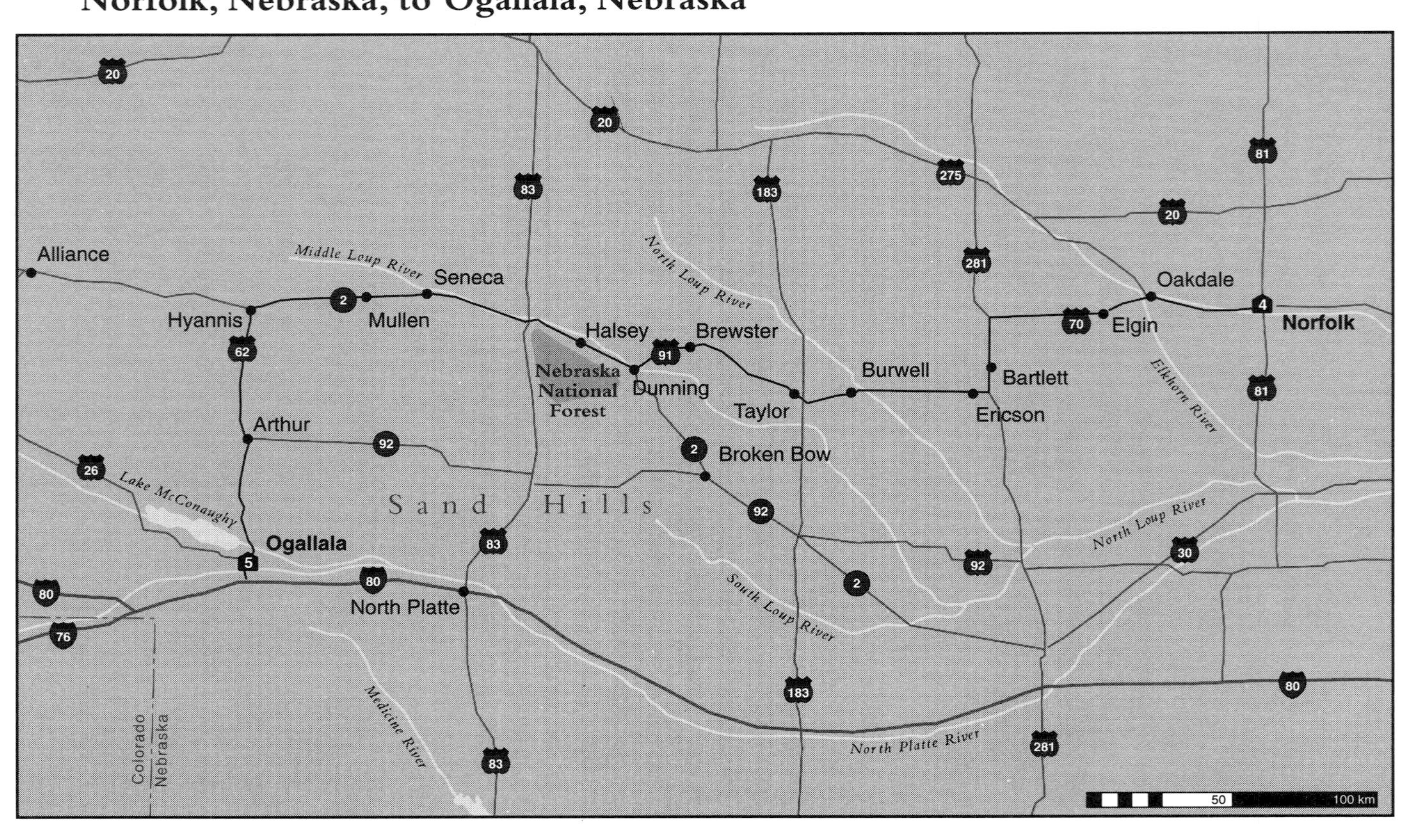

section. "Cowboys" on horseback, a Great Plains tradition, are fairly common sights in the larger cattle operations west of here.

At *Oakdale* (population 400), 28 miles west of Norfolk, leave Highway 275 and enter the town. The C. H. Brainard Lumber Company in Oakdale is a splendidly preserved example of a turn-of-the-century small-town business. Follow the unmarked secondary road west of the school on the southwest corner of Oakdale. The road soon begins to climb out of oak-forested margin of the Elkhorn River valley via a small tributary, Cedar Creek.

The road is climbing the Tertiary *Ogallala Formation.* At the top of the grade the transition to the High Plains geologic formations has been made. Although bedrock is mantled with wind-blown materials here as in the rest of central Nebraska, cedars are found on thin soils at the edge of the Ogallala. The name of Cedar Creek as well as the much larger Cedar River to the west marks a transition in vegetation. In the dry West, cedar and juniper typically occupy broken land.

The Ogallala Formation was produced by aggrading streams that deposited sediments eroded from the Rocky Mountains beginning about sixty-five million years ago. This formation is thus newer than the Rocky Mountains and was derived from them. The Ogallala Formation is the principal unit of the High Plains aquifer, a water-bearing formation now being extensively mined for irrigation purposes. It is the principal source of irrigation water from this point to eastern Colorado. Much of the expansion of irrigated agriculture on the High Plains in the past thirty years has been based on wells drilled into the Ogallala aquifer.

Turn right (west) at a T-intersection 7.5 miles south of Oakdale and proceed west between upland cornfields and pasture to the village of *Elgin* (population 730). Go straight ahead on Nebraska Highway 70 at the intersection of Highways 70 and 14. Past Elgin, the land surface begins to reflect the sparse look of the Great Plains. Pheasants are common here. Watch for them—they have a habit of casually walking on the highway. In summer, the prairie skies are alive with the song of the meadowlark. Upland plovers are common as well. A relict shelterbelt parallels the road for a distance after entering Wheeler County. The massive cottonwoods

Center-pivot irrigation has extended the raising of corn and other moisture-demanding crops onto the Great Plains. Even hilly surfaces can be traversed by the center-pivot's quarter-mile-long sprinkler system, which sweeps around the field on motor-driven wheels at the base of each support. Photograph by John C. Hudson.

in these human-planted rows of trees remain long after the farmsteads they sheltered from the wind have vanished. The higher elevations here have a sandy cover, a harbinger of the Sand Hills to the west. Signs for cattle ranches, with the rancher's identifying brand, also begin to appear.

Twenty-four miles west of Elgin, continue on Nebraska Highway 70 by turning left at its intersection with U.S. Highway 281 and proceed south to the village of *Bartlett* (population 140). To visit the town, turn right at the fairgrounds and drive west. Bartlett is a county seat whose old courthouse is now a public museum. The new courthouse is next door, a semi-underground block building.

Continue south on 281 about 7 miles, then turn right (west) on Nebraska Highway 70 to *Ericson* (population 132) in the *Cedar River* valley. Take Nebraska Highway 91 at its intersection with Highway 70, which is 17 miles west of Ericson. Seventeen miles west of Ericson the highway crosses the valley of Bean Creek, a tributary of the Cedar River. (Note: *Fort Hartsuff State Historical Park* is on a secondary road south from this point. The fort was built in 1874 and was abandoned in 1881, but nine original structures have been restored. It is open Memorial Day through Labor Day.) Cedar trees dot the slopes around Bean Creek. Today's cedars are relatively small, but those the early white settlers of the region found were enormous and they used them for buildings. The Great Plains was not as treeless as sometimes supposed.

After crossing the *North Loup River,* make a right turn to enter the business district of *Burwell* (population 1,280). Another county seat town, Burwell is laid out around a center square but its square contains business buildings rather than a seat of county government. The village was platted in 1884 when the railroad reached north into the Loup Valley.

Continue west from Burwell on Highway 91. A few miles west of Burwell, Highway 91 follows the ditches of the *Loup Valley Irrigation Project.* Droughts during the 1930s convinced many that irrigation was necessary in the Great Plains. At that time, before the invention of center-pivot sprinkler systems, which enabled upland irrigation from wells, irrigation was limited to valley bottoms where water could be diverted from a river. The U.S. Bureau of Reclamation (one of two major federal agencies concerned with dam construction in the West, the U.S. Army Corps of Engineers being the other) promoted diversion irrigation schemes such as this one in the Loup Valley. And they did so into the 1960s, after well-irrigation technology was available.

The irrigation ditches west of Burwell represent a late phase in the old irrigation technology and no new schemes of this sort are now under construction. But the controversy continues: ditch irrigation requires massive investment and floods large areas of would-be cropland for reservoirs; center-pivots drawing water from wells require no ditches, dams, or reservoirs, but they usually deplete the

The North Loup River meanders through a wild hay bottomland dotted with cottonwood trees near Taylor, Nebraska. Photograph by John C. Hudson.

groundwater supply faster than nature replenishes it. In wet years (such as 1991) corn grows well without irrigation, but betting on rain is risky in this zone of twenty inches of rainfall per year.

Continue on Highway 91 through *Taylor* (population 280). Taylor, another miniscule county seat in these small Sand Hills counties, is shaded by massive cottonwood trees. A log-cabin museum stands on the northwest corner of Taylor's center square.

Highway 91 west of Taylor follows the edge of the sandy uplands bordering the North Loup River. Twenty-five miles west of Taylor the highway again climbs above the river valley; the bluff here offers a good vantage from which to view the river as well as the typical Sand Hills countryside. The North Loup River meanders as it flows across the relatively new, flat surface between fringing dune formations. The land for miles north of the river is hay bottomland where wild hay is annually cut and baled for livestock's winter use. Cottonwoods dot the hay bottoms in the distance.

Thirty miles west of Taylor, enter *Brewster* (population 46), a still smaller county seat. Government services here are spartan, like everything else. About three-fourths of local taxes is used to support schools, although the county has a sheriff and other elected officials.

Presence in the Sand Hills of the stiff-spiked yucca, a desert plant that produces panicles of white, waxy flowers, may suggest that the climate is arid (yucca is New Mexico's state flower). In fact, the yucca's adaptability has more to do with soils than with climate in the Sand Hills. Sandy soils drain rapidly after a rain and therefore tend to be drier than fine-textured soils that receive an equal amount of precipitation. But rapid drainage means that Sand Hills soils do not have a calcium-accumulation horizon as close to the surface as other Great Plains soils in the same rainfall zone. The roots of tall grasses, such as big bluestem (*Andropogon furcatus*), can penetrate deep into sandy soil. The Sand Hills landscape thus includes the taller grasses more typical of the prairie to the east and also desert plants characteristic of truly arid environments found much farther to the west.

Dunning to Hyannis, Nebraska, 92 miles, Highway 2

West of Brewster the highway leaves the North Loup Valley and heads overland across the drainage divide through rolling dune topography. In 16 miles you will reach *Dunning* (population 182). Turn right (west) on Nebraska Highway 2 to continue through Sand Hills. From here to Hyannis along Highway 2 the route parallels the *Middle Loup River* and the Burlington Northern Railroad.

The railroad was built in the 1880s, the main object being to reach the Black Hills of South Dakota (rather than to serve the Sand Hills, which was then, as now, sparsely populated grazing country). The railroad brought some new settlers to the region who staked out land claims in the hay bottomlands along the branches of the Loup River. But passage of the Kincaid Act in 1904 allowed settlers to get 640 acres of free land in the Sand Hills and a tide of new would-be farmers arrived. Even 640 acres of land proved insufficient to support a family in this region because cultivation simply would not work. Removing the grass cover on sand almost immediately results in deflation (removal by wind) of the top soil layers, soon exposing sand surface. Evidence of the Kincaid era abuses in the Sand Hills largely disappeared once grassy vegetation became reestablished, although the presence of bare sand in road cuts and on old, unpaved road surfaces is a forceful reminder of the necessity for keeping the grass cover intact in this region.

One mile west of Halsey is the turnoff for the *Nebraska National Forest,* established in 1902 to provide settlers with a source of timber. Follow signs leading to the forest lookout tower. This forest is one of the nation's most remarkable. It is largely the creation of people, rather than nature, and reminds one of the long history of efforts to make forests in the grassland (another example was the Timber Culture Act of 1873, which gave settlers free land for planting part of it in trees). The road to the forest lookout tower passes extensive nursery grounds where pine seedlings are grown for distribution through the area. Note the mature forest fringing the older buildings next to the nursery. The road then climbs into the main area of ponderosa pine forest. On 5 May 1965, a lightning-caused wildfire destroyed nearly 11,000 acres of the forest; replanting commenced immediately. South of the parking area at the base of the fire lookout tower a trail leads through ponderosa pines and across rolling sand hills with prairie grass, yucca, and cactus among the trees.

Return to Highway 2 and continue west to *Thedford* (population 313), seat of Thomas County. (The Cowpoke Inn and Lounge in Thedford offers plain but tasty meals to local ranchers and to any tourists who happen this way.) At *Seneca* (population 90), 15 miles west of Thedford, set watches back one hour as you enter *Mountain Time Zone.*

A 110-car coal unit train, loaded in the Powder River Basin of Wyoming, moves east through a dry valley traversed by the Burlington Northern Railroad west of Mullen, Nebraska. Photograph by John C. Hudson.

Eleven miles west of Seneca is *Mullen* (population 554), the largest trade center in the central Sand Hills. The Hooker County courthouse just west of the business district in Mullen was built in 1912. Its seldom-used courtroom is straight out of the vintage West. The West seems more evident here in the Sand Hills than elsewhere, no doubt because the area never "progressed" beyond cattle ranching to crop agriculture. Ranchers driving pick-up trucks and hauling cattle in trailers to or from a weekly stock auction are the typical highway traffic.

The solitude of places such as Mullen is frequently broken these days, however, by the piercing air-horns of Burlington Northern coal trains. The railroad here amounted to little before the 1970s when the Wyoming coal boom got underway. Dozens of 110-car coal trains now leave the Powder River Basin of Wyoming every day, and a substantial number of them move through the Sand Hills on a much upgraded, computer-controlled railroad rebuilt to handle heavy traffic. Loaded trains run east to power plants in the Middle West, Oklahoma, and Texas; empty trains meet them heading west for Wyoming in what is sometimes a steady, two-way parade.

Highway 2 from Mullen to Hyannis 37 miles to the west closely follows the railroad through a series of apparent stream valleys that no longer contain streams. The east–west-trending dune ridges accumulated to spectacular heights in this section. Sand, apparently derived from local rock formations and stream beds, was produced during warm, dry interglacials of the Pleistocene and continued thereafter. Existing stream valleys were modified by wind action. Drainage was disrupted and numerous small lakes formed in scattered depressions. Dunes have continued to form and reform during the past several thousand years. As geologic formations, these are among the most recent to be seen.

Hyannis to Ogallala, Nebraska, 72 miles, Nebraska Highway 61

Hyannis (population 340) is the seat of Grant County, whose welcoming sign proclaims it as "The Best Cow Country in the World." Hyannis remains a viable, highway-oriented service center. Turn south about 1 mile east of Hyannis onto Nebraska Highway 61. This north–south highway, which crosses ridge after ridge of dune-formation hills, allows a different appreciation of the local topography from that experienced along Highway 2.

South of Hyannis the landscape is almost totally treeless. Here, as elsewhere in the Sand Hills, are numerous stock-watering tanks. Their pumps are almost silent and the only sound to be heard is the rhythmic squeak of the windmill; no electricity is necessary. Plenty of well water is available in the Sand Hills, which has a thick moisture-saturated layer beneath its surface. Note how the land surface around the watering tanks has been lowered, the result

Baled-hay house at Arthur, Nebraska. Baled hay was used experimentally as a building material in the Plains during the 1920s and 1930s. Tight bales of haystraw were secured by wooden laths driven into the bales, and the exterior was covered with stucco for durability. Photograph by John C. Hudson.

of cattle compacting the surface for years and removing the wind-shielding grass cover.

At *Arthur* (population 120), 34 miles south of Hyannis on Highway 61, can be found a collection of historic Sand Hills buildings, including a small log house along the highway near the north city limits. One-story, one- or two-room stucco and frame houses typical of dwellings on the sparsely settled, windswept Plains are found on every street in town. Arthur's greatests attractions are two baled-straw buildings which date from the 1920s. A church, two blocks west of the business district on Heath Street, and a house on Cedar Street two blocks north of the church were both constructed from tight bales of haystraw. Wooden laths were driven into the bales to strengthen the walls. Exterior and interior surfaces were then plastered. In the church (open to the public) a portion of the wall has been exposed to show the method of construction.

Continue south from Arthur on Highway 161. About 10 miles south of Arthur center-pivot irrigation fields begin to appear once again, a sign that one is leaving the Sand Hills. The rolling dune topography begins to flatten into longer, straighter slopes. About 20 miles south of Arthur the *North Platte River* valley is visible on the horizon. Soon after crossing another coal-hauling railroad line (Union Pacific line to northern Wyoming), Highway 61 passes the Martin Bay Recreation Area of *Lake C. W. McConaughy.*

Lake McConaughy, or "Big Mac" as it is known in western Nebraska, is a 30-mile-long reservoir on the North Platte River. Central Nebraska Public Power built Lake McConaughy in the late 1930s. (Note: Kingsley Dam was completed in 1941.) The project includes a small hydroelectric power-generating station. Lake water also is used for intake and cooling of coal-fired power plants. Lake McConaughy is a reservoir for irrigation water in this section of the North Platte Valley and its recreational opportunities have been important to the area.

Follow Highway 61 into *Ogallala* (population 5,095), a cow town in the days of long cattle drives north from Texas, now a livestock sales point and an interstate highway service center. Highway 61 crosses the Union Pacific Railroad's main line and, just beyond that, bridges low over the several braided channels of the *South Platte River.* Motels and restaurants are clustered around the Highway 61/I-80 interchange on the south side of Ogallala.

Before you retire, it is worth noting that the day began in Norfolk (elevation 1,527 feet) and closed in Ogallala (elevation 3,223 feet). Slowly, but surely, you are making your way to Denver, the Mile High City.

Day Five

THE HIGH PLAINS
Ogallala, Nebraska, to Greeley, Colorado, 228 miles

Travelers in a hurry to get to Denver—truckers, for instance—follow I-80 west from Omaha on the Nebraska-Iowa border and then branch to I-76 at Julesburg, Colorado, to make the final 180 miles. Both highways follow the Platte and South Platte rivers. Except in a few Colorado segments where the highway leaves the river, the entire routes is so tree-fringed or irrigated that one gets little sense of the Great Plains environment and how it differs from that of the Middle West.

Day Five of the journey seeks to remedy that by crossing the High Plains surface itself rather than by threading the region's major river valley. The day begins by driving south from Ogallala into a major irrigation district of southwestern Nebraska, then heading straight west across the wheat lands of Colorado to reach Greeley on the irrigated Colorado Piedmont.

Ogallala to Imperial, Nebraska, 47 miles, Highway 61

Drive south from Ogallala on Highway 61. About 3 miles south of the I-80 crossing, Highway 61 climbs the side of the South Platte River valley. Dryland wheatfields begin to replace irrigated cornfields. This is one segment of the largest wheat-raising area in the

Ogallala, Nebraska, to Greeley, Colorado

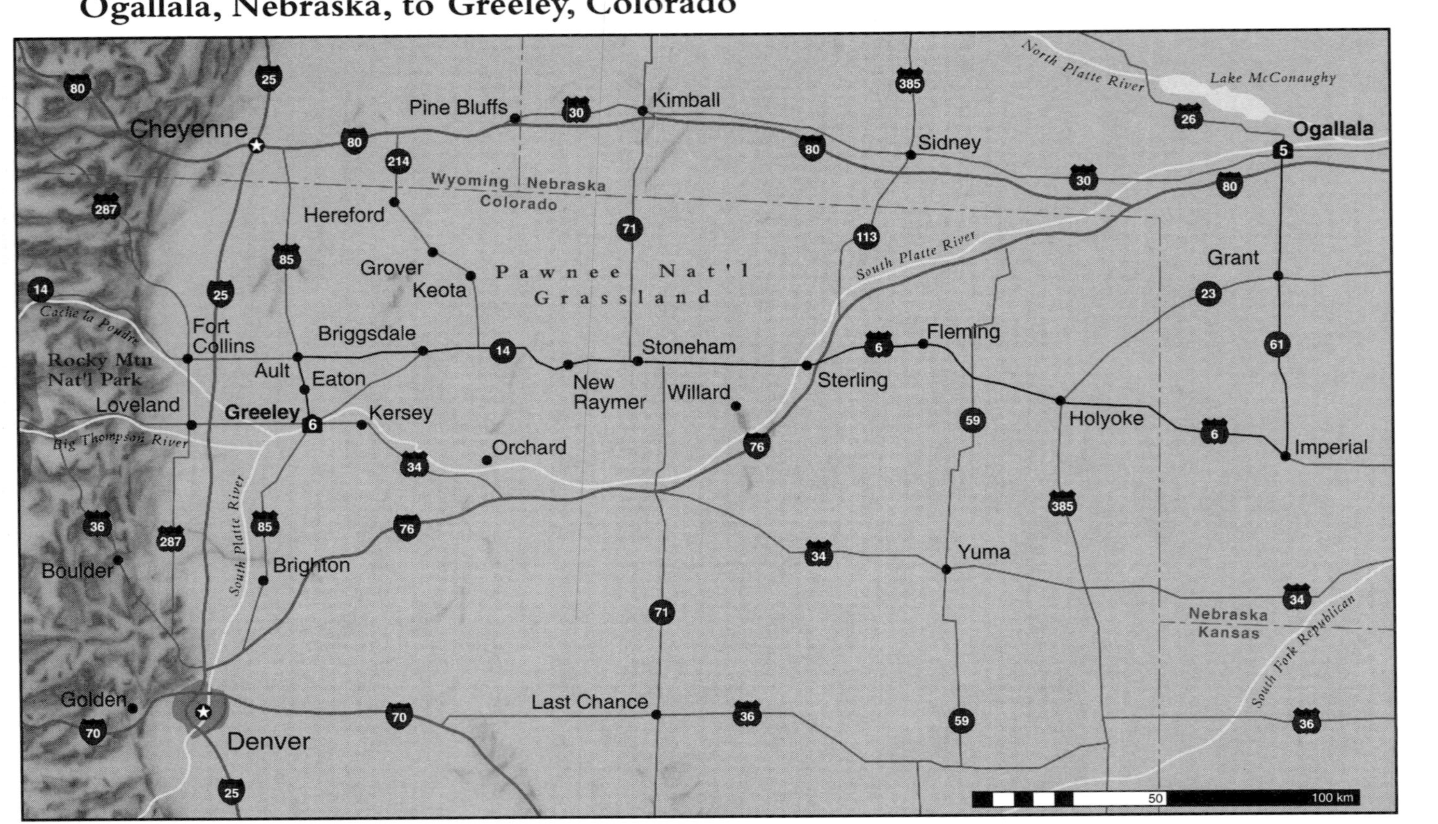

Dry-farmed strips of winter wheat alternate with fallow strips on the High Plains of eastern Colorado. The fields are oriented to minimize wind erosion. The underlying pattern of elongated dunes, produced by winds with the same direction as today, is visible as a streaked pattern across the striped fields. Photograph by John C. Hudson.

United States—the winter wheat belt, which stretches from Nebraska to Texas. Winter wheat, which is sown in late summer, remains in the ground over the winter and, thus, is ready to resume growth once warm weather arrives the following spring. The crop is harvested in early to mid-summer. Winter wheat is the principal grain used for milling bread flour. The central Great Plains region produces more than twice as much wheat as the domestic market demands. Between half and two-thirds of the crop is exported and most of the export flow is south to Texas Gulf ports.

The usual culture for winter wheat in this zone of semi-aridity is to use half the land each year for the crop and let the other half lie fallow; in other words, to use two years' precipitation to grow one year's crop. Dry farming, as this method is known, is a means of conserving soil moisture, but it also demands that fallow fields be cultivated to remove unwanted, moisture-robbing weeds. Planted strips alternate with fallow strips, creating a spectacular view from an airplane (although difficult to appreciate on the ground).

Follow 61 south until you reach *Grant* (population 1,240), about 19 miles south of I-80. The south end of the business district (Central Avenue) is marked by massive grain-storage elevators—cylindrical concrete structures that are typically used to store wheat. Most wheat farmers in this region store little harvested grain on their farms. They bring the crop to elevators in town as it is harvested, and the result is that even miniscule towns often have massive grain-storage facilities.

Just south of Grant, Highway 61 (which joins State Highway 23 east for 3 miles) passes more agribusiness operations, including a fertilizer terminal and a bean-processing company (dry, edible beans such as pinto and Great Northern varieties). Irrigation begins to reappear in this section. The spikes of yucca on rolling hills indicate that the landforms are sand dunes. These dunes, like those in the Nebraska Sand Hills, are fairly recent features, formed between 3,000 and 1,500 years ago during a period of climatic aridity.

The Ogallala aquifer underlies this section also, and where the surface is not too sandy the land is now used for irrigated crops. Corn demands more moisture than wheat, but, with sprinkler irrigation to supply the moisture, this southwestern corner of Nebraska is now undeniably part of the Corn Belt. Corn brings a higher price per bushel than wheat and most farmers grow both, although they typically do not irrigate the less remunerative wheat crop. As Highway 61 nears Imperial, 25 miles south of the junction of 61 with 23, center-pivot–irrigated cornfields dominate the scene.

Follow Highway 61 to its junction with U.S. Highway 6, then turn right (west) and enter *Imperial* (population 2,010). Although appearances may not suggest it, this town is an important link in the international trade of the United States. The rows of grain silos here have a capacity that allows loading 5,000-ton trainloads of corn—an unremarkable fact today, except that corn is produced in large quantities in very few areas west of here. Imperial is at the fringe of the Corn Belt, but it is closer to the Pacific Coast than any other corn-producing region in North America. And given the increasing demand for corn and other feed grains in the Pacific Rim countries, southwestern Nebraska is now in the front line of international supply. Local feedlots consume whatever corn is not shipped out.

Imperial to Sterling, Colorado, 85 miles, Highway 6

Highway 6 passes through more miles of irrigated cornfields west of Imperial. Note that the outside corners of square fields not reached by the circular sweep of irrigation are planted in wheat, thus making maximum use of the land. Irrigation has brought weed problems to this section of the Plains. Only fifty years ago this was the margin of the Dust Bowl, where neither crops nor weeds would grow. But irrigation since the 1960s has allowed exotic weeds to thrive. Herbicide application is the typical means of coping with weeds, and hence agricultural chemicals have come into widespread use. The expansion of irrigated cornfields into the old Wheat Belt has had many consequences.

The land slopes perceptibly upward to the west as we cross the Colorado state line 24 miles west of Imperial. The slope is formed on the underlying Ogallala formation, which increases in elevation about sixteen feet per mile in this section. Grain elevators visible to the north of Highway 6 are located eight miles away in Amherst, Colorado.

At *Holyoke* (population 1,930), a county seat and trade center, wheat elevators again dominate the town skyline. Continuing west, note the large feedlot just east of the limits of *Haxtun* (population 950). This lot feeds approximately 4,000 head of beef cattle, which are sold to meat packers in Sterling or Greeley. The land around the feedlot is used to grow corn, although this feedlot owner, like most, buys corn to supplement the owner's crop. The most obvious environmental problem posed by a feedlot is its aroma of rotting corn (rather than livestock manure); most farmhouses near feedlots have understandably been abandoned. Chemically rich feedlot wastes require holding ponds and treatment before they can be released. The all-weather ponds sprout exotic plant life brought by migrating Canada geese, which find the ponds attractive stopping points. The nauseating smell and pitiful sight of thousands of cattle standing passively and quietly in the mud and manure are enough to make some people postpone their next T-bone steak indefinitely. But this is how most of our beef is produced today.

Just west of *Fleming* (population 345) a maximum elevation of 4,320 feet is reached before Highway 6 slowly descends the South Platte River valley to enter *Sterling* (population 10,360), which is 400 feet lower than the High Plains to the east. Irrigated land in eastern Colorado has long been an important producer of sugar beets. The Great Western Sugar Company has sugar refineries in Sterling as well as in Fort Morgan (population 9,100) and other nearby cities. Most sugar-beet plants operate only seasonally and their abandoned look can be misleading.

Sterling to Ault, Colorado, 83 miles, Highway 14

Proceed west through the business district of Sterling to reach Colorado Highway 14 (West Main Street), the city's fast-food strip. Highway 14 begins a long climb out of the South Platte Valley past irrigation ditches. The South Platte lowland is on the (Cretaceous) Pierre Shale, whereas the bluff lines on either side of the lowland are formed on the (Upper Cretaceous) Fox Hills sandstone. The (Oligocene) White River chalk lies atop the Fox Hills Formation. Passing occasional pumping oil wells, Highway 14 crosses Pawnee Creek and climbs to the level of the Tertiary formations once more. Thirteen miles west of Sterling a secondary road (with a sign pointing south to Willard) crosses Highway 14. Travel a few miles either north or south of Highway 14 on this cross-road to get a good view of the sharp edge of the Fox Hills plateau. Prairie-dog "towns"—hundreds of burrows made by these sleek rodents of the grassland—account for the low earth mounds on the landscape. Deer and jackrabbits are commonly sighted here as well.

Once out of the South Platte Valley, dry-farmed winter wheat reappears as the dominant land use. The chalky buttes visible in the distance north of Highway 14, 4 miles east of *Stoneham* (unincorporated), are remnants of the easily eroded White River Formation. Stoneham, as well as New Raymer, Buckingham (not on your Rand McNally road atlas), and Keota to the west were railroad-created towns along a line built to Cheyenne, Wyoming, that served Colorado wheat country. The railroad has been removed and the towns have long lost their roles as trade centers. Most of the farmsteads have vanished as well, but the wheat crops—some the work of absentee "suitcase farmers" from nearby areas—remain. *New Raymer* (unincorporated) is a virtual museum of 1920s-era store buildings, all abandoned but reasonably intact. U.S. Air Force Minuteman-III missile silos are located near the highway beginning at Stoneham, and working oil wells reappear west of New Raymer. This is a productive land with an odd assortment of uses, but it is no longer a place people want to live.

The *Pawnee National Grassland* occurs in large scattered tracts to the north. Some of these acres were cultivated for wheat and then abandoned when they produced poor crops and the steady west wind removed the topsoil. The federal government created the grassland reserve in 1937 (Bankhead Jones Act) through purchase of marginal tracts, and the U.S. Forest Service assumed responsibility for its management from the Soil Conservation Service in 1954. The pronghorn, an antelope-like ruminant, is common here, usually in groups of six to ten bounding across the open grass/shrub range.

Ten miles west of New Raymer a sign pointing to Keota, Grover, and Hereford indicates a gravel road that leaves Highway 14 to the northwest. (Note: unpaved roads are frequently impassable here in wet weather when the dust turns to slippery, sticky clay.) If conditions permit safe travel on this gravel road, then visit *Keota* (population 4), 5.5 miles north of Highway 14. The smallest incorporated place in the state of Colorado, its buildings have completed their second adaptive reuse and are now abandoned forever; two house-trailers remain. The municipal water tower still stands sentinel above the unused streets of this town, which had a productive life of only a few decades before it began to fade away. Keota was fictionalized as "Line Camp" in James Michener's novel *Centennial* (1974), and it has been a favorite of photographers searching for a ghost town on the Plains. Nineteen miles northeast of Keota via another gravel road are the *Pawnee Buttes,* elevation 5,375 to 5,500 feet, which are capped by the erosion-resistant (Miocene) Arikaree Formation.

After returning from Keota south to Highway 14, turn right (west). Note the dry-farmed wheat and fallow strips which here are oriented on a northeast–southwest diagonal, ninety degrees to the prevailing wind direction—yet another attempt to work within the limits the environment has set. Two miles northwest of *Briggsdale* (unincorporated) is a small recreation area that is part of the Pawnee National Grassland. On clear days, peaks of the Rocky Mountain Front Range may be glimpsed on the high ground this far east of the mountains. Continue on 14 west to Ault.

Ault to Greeley, Colorado, 13 miles, Highway 85

As the long-awaited view of the Rocky Mountains sharpens against the western horizon, Highway 14 descends in elevation to the level of the *Colorado Piedmont* a few miles east of *Ault* (population 1,110). Turn left (south) on Highway 85 at Ault, typical of small agricultural processing centers in the region. The term "Colorado Piedmont" refers to the area between the Rocky Mountains on the west and the edge of the Tertiary formations, such as the Ogallala, on the east. On the Piedmont the Tertiary layers have been removed by uplift and erosion. Unlike other aquifers beneath it, which outcrop in the Rocky Mountain foothills, the Ogallala cannot be recharged by water from the base of the Rocky Mountains because of the missing section across the Piedmont. This is also

Feedlot along the South Platte River east of Greeley, Colorado. This is one of the largest beef cattle-producing areas of the United States today and the area is increasingly becoming important in meat packing. Photograph by John C. Hudson.

why the Ogallala is often said to contain "fossil water." Its future rate of productivity will eventually have to decline.

The east-facing slopes of the Rockies are drier than those facing west, and for the past five decades irrigation water for the Piedmont has been transferred from the Colorado River Basin. Water is pumped up to Grand Lake, a reservoir in Rocky Mountain National Park, and from there is conveyed via a thirteen-mile tunnel through the Front Range to the Big Thompson River, a tributary of the South Platte. It was an early case of stealing water from one drainage basin to supply another, although in the late 1930s, when the *Colorado–Big Thompson Project* was launched, demands on the Colorado River's water were a fraction of what they are today. The Piedmont became a vast, productive plain of irrigation farms

SIDE TRIP TO THE GREELEY-AREA FEEDLOTS

The Colorado Piedmont around Greeley is one of the most intensive livestock-feeding areas in the United States. Follow Colorado Highway 263 (8th Street in Greeley) east from its junction with the U.S. Highway 85 bypass on the east side of the city. Large feedlots are to be seen along this portion of Highway 263, which follows the South Platte River. At the junction of Highways 263 and 37, turn right (south) on Colorado Highway 37 to *Kersey* (population 980). In Kersey turn left (east) on U.S. Highway 34. A few miles east of Kersey is the Monfort feedlot, the largest in the United States. It is impressive for its sheer size, although the smaller lots along Highway 263 (to the north of 34) offer a better close-up of operations.

The feedlots and irrigated stock farms of the South Platte River valley can be followed another 30 miles east of Kersey to reach the town of *Orchard* (unincorporated), which was the prototype for the town of "Centennial" in James Michener's celebrated novel. (Note: A few miles east of Masters on Highway 34, turn left (north) on State Highway 144 and cross the South Platte River to reach Orchard.) Orchard's revival, based on the popularity of the book's serialization for television, seems to have been short-lived. Many of the town's buildings remain, however, and the various attempts at preservation have captured the feel of a late-nineteenth-century railroad town on the Plains.

Return to Greeley via U.S. Highway 34.

along the South Platte as a result of important Colorado River water.

The western edge of the Colorado Piedmont is part of the *Denver Basin,* a deep downwarp of rock strata at the edge of the Rocky Mountains. Looking toward the mountains one notices the land surface sloping downward before the foothills are reached. The Piedmont here is a maze of irrigation ditches, laterals, and drains. Crops of sugar beets, onions, corn, and alfalfa alternate in the patchwork of small fields. Many contrasts are evident coming from the dry-farmed uplands to the irrigated lowland: feedlots reappear, farms get smaller, population density increases a hundredfold, agricultural chemicals are in widespread use, weeds and trees replace grasses, and the typical resident is somebody's employee rather than a farm owner.

Follow U.S. 85 south to *Greeley* (population 53,000), principal city of the agricultural Piedmont. In 1870 Horace Greeley's Union Colonists arrived here and began diverting water from the *Cache La Poudre River* to irrigate their farms. On the north side of Greeley is the Monfort Company packing plant (now a subsidiary of Con-Agra), which slaughters animals from its large feedlot east of Greeley. The broad streets of downtown Greeley are typical of those found in western towns, although the attempt to turn the business district into a tree-shaded mall is not.

Greeley, Colorado, to Denver, Colorado

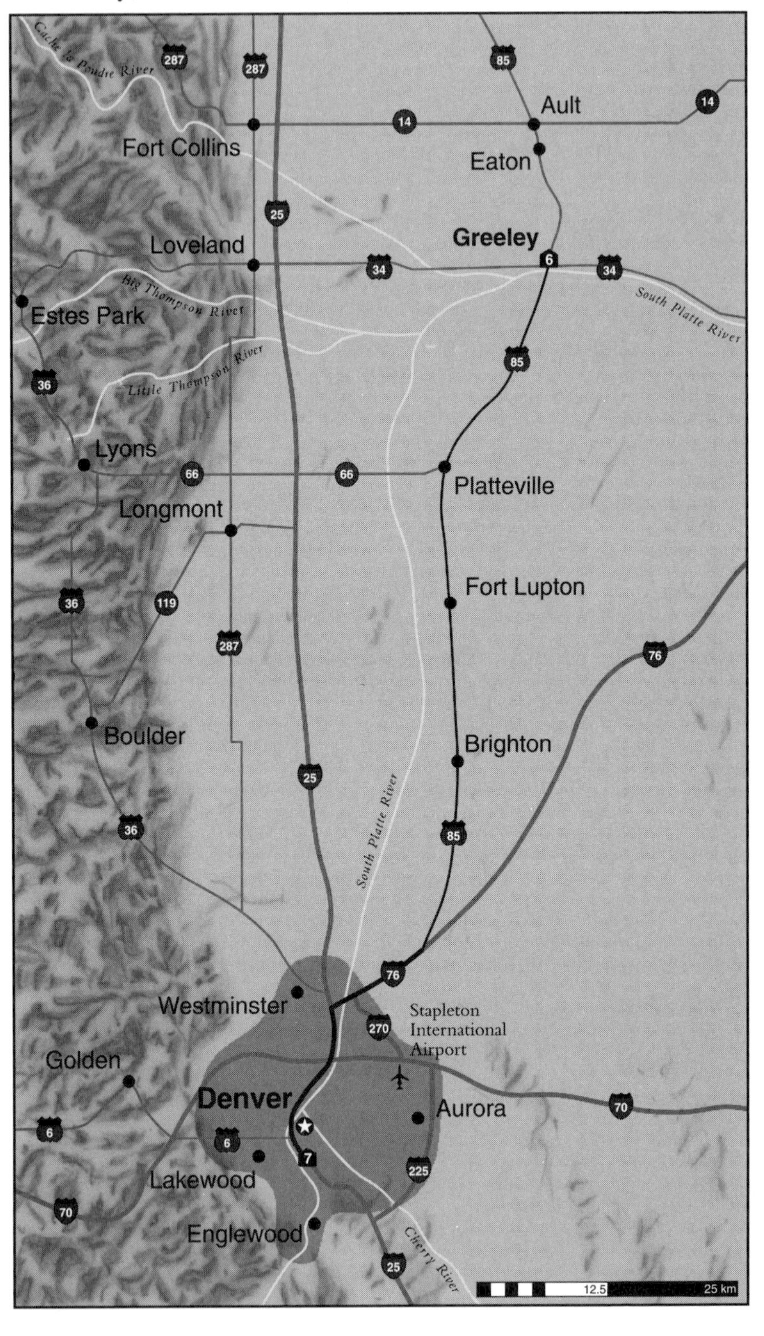

Day Six

DENVER AND THE COLORADO PIEDMONT
Greeley to Denver, Colorado, 54 miles

This is a short day in terms of travel, but one that offers some time to explore the city of Denver (and, if the shoe fits, a break in "The Transcontinental" journey).

Greeley to Brighton, Colorado, 30 miles, Highway 85

Leave Greeley (elevation 4,664 feet) by following U.S. Highway 85 south through *Evans* (population 5,880), a highway-oriented strip town. Highway and local traffic build steadily as we near Denver. Irrigated fields of the Colorado Piedmont are here set against the dramatic backdrop of the Front Range mountains 40 miles to the west. Fields of onions, potatoes, corn, alfalfa, and sugar beets line the highway. Elevation increases toward the south as the highway follows upstream along the South Platte River, which, in turn, is following the trough of the Denver Basin. Continue on 85 to Brighton, about 30 miles south of Evans.

The local population acquired a strong Hispanic component as a result of irrigation developments since the 1930s that demanded increasing amounts of agricultural labor. Mexico was the principal source of field workers for the Piedmont as well as for other irrigated sections in the West. Although some Hispanic agricultural workers are still seasonally migratory, most reside permanently in small settlement clusters along field margins. Denver, and all the Piedmont cities, have large Hispanic populations today.

Brighton to Denver, Colorado, 24 miles, I-76 and I-25

South of *Brighton* (population 14,200) follow the signs to I-76 (exit 12) leading south. The skyscrapers of *Denver* (population 467,600) are visible as far north as exit 9 on I-76. The highway leads through grain-storage terminals, railyards, and oil refineries on the northern outskirts of the urbanized area. Continue following highway signs indicating "Denver," which lead via I-25 into the city's heart. (Note: I-25 is congested and traffic is slowed by nearly constant construction projects that are attempting to increase the capacity of this single, limited-access highway into Denver.) The panorama of the Denver skyline continues south along the South Platte River. Leave I-25 at exit 212B, then follow city streets north through one of Denver's Hispanic neighborhoods to 16th Street. Follow 16th Street east across its viaduct over the South Platte River and past Denver's Union Station.

Gold was discovered near Denver in the 1850s. The city emerged from the collection of mining camps surrounding the gold strike, and by 1880 the city had 35,000 inhabitants. Railroads arrived from the east in the 1870s and more lines of track soon twisted up through canyons and passes into the mountains, but Denver lacked transcontinental status until the Moffat Tunnel, through the Front Range, was opened in 1927. The mountains west of Denver have some of the most difficult eastern approaches of any in the Rockies; transcontinental railroads chose easier routes, across Wyoming to the north or across New Mexico to the south. In Colorado, only the narrow defile known as the Royal Gorge (west of Canon City) leading to remote Tennessee Pass (elevation 10,424 feet) north of Leadville offered a plausible route for a transcontinental railroad.

Despite being somewhat off the path, Denver grew from its dual role as a major city of both the Western Great Plains and the Rocky Mountains. Mineral smelting remained a major source of employment into the twentieth century and the city's stockyards were among the largest in the West. As a wholesaling and trading center, Denver was the focal point of regional commerce. Parts of its colorful commercial past have been preserved along *Blake Street,* at the end of the Sixteenth Street viaduct. Other historic districts, including the *14th Street Lower Downtown Denver Historic District, Market Street,* and *Larimer Street,* are in the same neighborhood. They escaped the wrecking ball of urban renewal, common in the nation's cities between the 1950s and 1970s, and today offer the visitor a variety of interesting shops and restaurants.

Denver's core of downtown skyscrapers mirrors its more recent economic history. *Seventeenth Street* is a skyscraper canyon between Market Street and Broadway. By the 1950s, the Piedmont and Rocky Mountain Foothills region were attracting white-collar employment and research-oriented firms from all over the United States. The amenities-based migration of the 1950s and 1960s made the Denver–Boulder area a sort of scaled-down version of the aerospace growth areas of California. By 1970 Denver had half a million people. High-rise downtown office buildings reflected the city's new economic fortunes in oil, computers, cable television, and other telecommunications industries.

The 1970s oil boom turned to a bust in the 1980s and that was followed by layoffs in the high-technology electronics and defense industries and then by failure of several important financial institutions. By then it was clear that Denver had overbuilt. Today Denver has reached a sort of equilibrium between cycles of growth and decline, although its future growth seems assured given the city's reputation as a good place to live and to do business.

Turn right (south) on Broadway at Seventeenth Street to reach the *Denver Civic Center,* which includes the Colorado State Capitol, Denver Mint, and Colorado Heritage Center (public museum devoted to the state's history). The Colorado Historical Society's

Byers-Evans House, at Thirteenth and Bannock streets displays Denver's history. In the same Civic Center area is the *Denver Art Museum,* which has a fine collection of western art that is very much worth seeing.

BEYOND THE GREAT DIVIDE

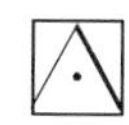

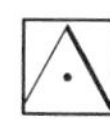

Crossing the Heartland traversed two great regions of the United States—the western half of the Middle West and the Great Plains. You are now in Denver, Colorado, at the western edge of the Great Plains and at the gateway to the West. Here is where The Transcontinental continues on its third leg of a great sojourn across the United States. It began in Washington, D.C., crossed the Appalachians, proceeded to Lake Michigan and Chicago, and then crossed the Heartland.

The third leg of The Transcontinental is called *Beyond the Great Divide,* and it, too, takes the traveler across two great regions of the continent—the Southern Rocky Mountains and the Colorado Plateau. Cotton Mather, P. P. Karan, and George F. Thompson, your guides on this next segment, will take you to exceptional places and show you how ancient cultures and peoples are still alive and well in the modern world. There is a "tableau without parallel," but before you entertain the thought of that journey, let me bid farewell.

PART THREE

Resources

Hints to the Traveler

If you are unfamiliar with the Heartland, you might as well think of it as a foreign country and conduct yourself here as you would in any place about which you have some uncertainty. But I can provide a few hints that will make your trip across the mid-continent region even more pleasurable than if you arrived unprepared.

CLOTHING

Casual dress is appropriate all year, everywhere, in the Middle West and Great Plains. Hot-weather clothing is necessary from June through August and, occasionally, in May and September. Clothing comfortable in freezing temperatures is necessary from mid-October through mid-April. Above all, be prepared for changes in the weather! Days can be cool in the warm season, especially under rainy conditions. Likely as not, such days will be preceded and followed by hot, sunny conditions. This is a mid-continent location and, hence, one can expect anything from a gentle breeze to a strong wind during the daylight hours, increasingly so with distance toward the west.

FOOD

Cities of this region contain the same mix of fast-food establishments found everywhere else in the United States. Even in the sparsely settled Great Plains, one can find the Golden Arches at many interstate highway interchanges. Do not expect to find the usual franchise-food outlets in small towns that do not have major highway, however.

Away from the cities and interstates, eating establishments still tend to be one-of-a-kind cafes. The food is often good, though seldom outstanding. In general it is wisest to order the simplest and most frequently cooked items on the menu. Steaks and hamburgers are available everywhere. Roast beef and ham are usual supper (evening meal) and Sunday dinner entrees. Deep-fat fried foods, such as chicken, french fries, fish, or pork tenderloin, are usually a good bet. In this part of the world, a "salad" means a small plate of shredded head lettuce to which a sweet, sticky, orangish-colored liquid has been applied sparingly. "Toast" means a square slice of spongy white bread, toasted. Many cafe owners pride themselves on the pies they offer. A piece of blueberry, peach, or banana cream pie is often the best part of the meal. "Home cooking" is sometimes advertised; this refers to the type of meal once served in homes (meat, potatoes, vegetables, dessert) and does not mean that the food was cooked in somebody's home.

A great American myth, which can be tested here, concerns the high quality of food available at truck stops. While the amount of food on the plate is usually greater at a truck stop, quality rarely matches quantity. Truck drivers eat the same food, day after day, often in the same truck stops, and probably take less interest in what they eat than just about anyone else you could meet.

One way to pick the best restaurant in a town (if there is a choice) is to select the one that looks busiest, has the most farmers' trucks parked in front, and appears to be the cleanest. It is hard to go wrong at breakfast or lunch. A tasty evening meal is harder to come by, however, because this is the time of day when most local people have finished work and are at home; hence, the cafes have fewer customers. Some close at 1700.

MOTELS

All establishments having overnight accommodations are called motels. Quaint, cozy "bed and breakfasts" are not of the Corn Belt or Great Plains. If you are in a remote area and need a place to stay for the night, the best strategy is to travel to the nearest interstate-highway town, which is where the motels will be found. Small-town motels can be very good even though they appear in no guidebooks or directories and are not affiliated with a major chain. They are sometimes booked ahead, especially on weekdays, by salesmen, construction crews, and others living temporarily away from home. Sometimes all the motel space in a town is taken up by the relatives attending someone's wedding, or by itinerants working that week at the local county fair. If you must stay in a particular place, then reservations are recommended.

SERVICES

It is best to purchase gasoline anywhere there is another service station nearby so as to avoid paying an uncompetitive price. Most of these establishments still perform the "service" with which they are identified (check the oil, clean the windshield, check the tire pressure, and the like). They are invariably run by people who know how to deal with mechanical problems. (Note: Parts for foreign cars, however, can be a problem anywhere in rural areas.) Those in need of a restroom, but not in need of gasoline, should keep in mind that every county has a courthouse (usually in the middle of the county's largest town), every courthouse has restrooms (usually in the basement), they are clean, and they are free and open to the public.

If you have a breakdown or encounter other emergencies en route, you will discover that the people of this region will be unselfish and unhurried in giving you assistance. Do not be afraid to offer them a monetary reward for their help.

The main-traveled roads of mid-America see a small but constant traffic of assorted folks determined to cross the country as cheaply as possible. As elsewhere in the United States, be cautious about accepting hitchhikers.

And, finally, on a related matter, for you want to avoid the service of a lawyer, the enforcement of highway speed limits is variable today, and it is conducted from the air as well as on the ground. Some state and local governments fine speeding simply as a source of revenue while others are determined to get traffic moving more slowly. Getting caught exceeding the speed limit by more than 5 miles per hour is likely to result in a fine.

WEATHER

When to Travel

Early June through mid-October is the best travel season, though some travelers extend the season a little earlier to include mid-May. The countryside takes on the fresh look of a new crop by late May. The wild grasses are green through June, then they slowly begin to turn a sage color. Corn and soybean fields are green until early September, but then begin to wither. The wheat fields of the High Plains turn amber by July. Wildflowers are most abundant in June, when the roadsides can be quite beautiful

Summer days are hot, generally in the mid-80s (30°C), and nights cool to the low 70s (22°C). Higher temperatures in the Plains are offset by a corresponding decrease in humidity. While 90° is uncomfortable in Chicago, it is much less so in western Nebraska. Hot days continue through September, although overnight low temperatures begin to decrease at that time of year. Late summer and early fall are excellent for travel: the weather is typically sunny, nights are cool, and the air is dry.

When Not to Travel

Winter in the Heartland is severe. The proverbial "blizzard" is the most crippling winter storm because it is accompanied by high winds and rapidly falling temperatures. Blizzards are rare, but should be taken extremely seriously. No one should attempt to travel during a blizzard. In fact, anyone contemplating making this journey from Lake Michigan to Denver should think twice about it after mid-November or before mid-April. Most snow storms are fairly light and highway crews get the roads plowed quickly after a snowstorm. But blowing snow and icing on the roadway persist long after a storm. Travel across the Great Plains by automobile during winter, especially late winter, is not recommended even on the interstate highways. Winter road closings are so common that many freeway interchanges are equipped with gates that prevent access during bad weather.

Spring is a brief season. It usually arrives without notice and is marked by a rapid melting of snow and hot daytime temperatures. This can occur at any time from mid-March to mid-April, but the transition is never complete until later. Hot days at that time of the year can be a prelude to snow the next. Snow can occur in eastern Colorado and the Piedmont through May, although the storms are brief and snow and ice melt almost immediately. Early spring is a risky season for travel, but it is also an uninteresting time of year because the fields are either bare or covered with dead vegetation and the ground is soggy.

Watching the Weather

Weather is a subject of intense interest for obvious reasons. Local radio stations repeat weather forecasts constantly during the growing season, just as often as they update the price of corn futures at Chicago or live hogs at Peoria. Those two kinds of news are related, of course. Bad weather that suggests a poor harvest will be anticipated in the grain futures market, just as a large crop will drive down the price of corn fed to livestock. Bets on the weather tend to drive bets on the prices of agricultural commodities several months in the future. It is an endless game, observed closely by every farmer, grain dealer, and banker.

While the mid-continent receives most of its precipitation during the summer months, there is no rainy season that can be planned on (or around). Rare is the year that has a dry spring, summer, *and* fall. The more usual case is for one or two seasons to be dry, another one or two to be wetter than average. Weather patterns typically persist for a time: dry weather is followed by more dry weather; cloudy or rainy days similarly tend to bunch together for several days, sometimes several weeks. Another pattern is a succession of small atmospheric disturbances that may pass through every few days, so that the weather alternates from rain, to clearing, cooling, warming, cloudy, and then back to rain. All of which is to say that the weather is highly variable, although not unpredictable.

Those who travel across the breadth of the mid-continent naturally worry about tornadoes. The Lake Michigan-to-Denver traverse described in this field guide lies north of "Tornado Alley," the belt where violent storms are especially common. Tornado Alley stretches from the Texas Panhandle across Oklahoma to Kansas. Squall lines—bands of thunderstorms that typically move north-northeasterly in summer and are accompanied by lightning and brief, heavy rains—are the breeding grounds for tornado conditions and for this reason they are monitored constantly. Severe storm warnings are frequently issued in late afternoon or early evening on hot, humid summer days when these squall lines gather intensity. Their passage overhead typically lasts between thirty minutes and one hour. Few of them are life-endangering, however.

During your trip across the Heartland, pay attention to the weather no matter what the time of year. Monitor weather forecasts as necessary over the local radio stations and, if your motel offers cable television, watch the Weather Channel during the evening for the region's "next-day" forecast.

Suggested Readings

For regional geography the standard works remain Nevin M. Fenneman, *Physiography of Eastern United States* (McGraw-Hill, 1938) and *Physiography of Western United States* (McGraw-Hill, 1931). Also useful are William D. Thornbury, *Regional Geomorphology of the United States* (John Wiley, 1965); and Charles B. Hunt, *Natural Regions of the United States and Canada* (Freeman, 1973). More advanced, but nonetheless readable, is Robert V. Ruhe, *Quaternary Landscapes in Iowa* (Iowa State University Press, 1969). The classic description of regional climate is John R. Borchert, "The Climate of the Central North American Grassland," *Annals,* 40 (1950), pages 1–39. (Note: *Annals* refers to *Annals of the Association of American Geographers,* a quarterly publication.) Benchmark surveys of the professional literature are found in two valuable collections: H. E. Wright, Jr., and David G. Frey, *The Quaternary of the United States* (Princeton University Press, 1965); and H. E. Wright, Jr., editor, *Late Quaternary Environments of the United States,* two volumes (University of Minnesota Press, 1983).

The "frontier theory" of settlement, which has influenced much of the scholarship about the mid-continent, can be found in Ray Allen Billington, *Frontier and Section: Selected Essays of Frederick Jackson Turner* (Prentice-Hall, 1961). Comprehensive studies of the Middle West include James H. Madison, editor, *Heartland: Comparative Histories of the Midwestern States* (Indiana University Press, 1988); and James R. Shortridge, *The Middle West: Its Meaning in American Culture* (University of Kansas Press, 1989). Population sources are discussed in two papers by John C. Hudson, "North American Origins of Middlewestern Frontier Populations," *Annals,* 78 (1988), pages 395–413; and "Who Was 'Forest Man'?: Sources of Migration to the Plains," *Great Plains Quarterly,* 6 (1986), pages 69–83. The most enduring monograph on any region of the United States is Walter Prescott Webb, *The Great Plains*

(Ginn and Company, 1931). James A. Michener's novel *Centennial* (Random House, 1974) is also worth reading. An American Indian perspective on the Plains is offered in another classic, Dee Brown, *Bury My Heart at Wounded Knee: An Indian History of the American West* (Holt, Rinehart & Winston, 1971). Also see E. Cotton Mather, "The American Great Plains," *Annals,* 62 (1972), pages 237–257; and John Fraser Hart, "The Middle West," *Annals,* 62 (1972), pages 258–282.

Land survey systems are described in Hildegard Binder Johnson's *Order Upon the Land: The U.S. Rectangular Land Survey and the Upper Mississippi Country* (Oxford University Press, 1976). The standard agricultural history of the Middle West is Allan G. Bogue, *From Prairie to Corn Belt: Farming on the Illinois and Iowa Prairies in the Nineteenth Century* (University of Chicago Press, 1963). For a modern, geographical perspective on the subject see John Fraser Hart, *The Land That Feeds Us* (W. W. Norton, 1991). The culture of farming on the High Plains is explored in Leslie Hewes, *The Suitcase Farming Frontier* (University of Nebraska Press, 1973).

Chicago's role as the metropolis of the Middle West is developed in William Cronin's *Nature's Metropolis: Chicago in the Great West* (W. W. Norton, 1991). Chicago's geography is described in Harold Meyer and Richard Wade, *Chicago: Growth of a Metropolis* (University of Chicago Press, 1969); and Irving Cutler, *Chicago: Metropolis of the Mid-Continent,* second edition (Kendall-Hunt, 1976); an excellent update of these two standard sources is Michael P. Conzen, "The Changing Character of Metropolitan Chicago," *Journal of Geography,* 85 (1986), pages 224–236. Stephen J. Leonard and Thomas J. Noel, *Denver, Mining Camp to Metropolis* (Colorado Associated University Press, 1990), is a comprehensive history of that city. Two books by John W. Reps, *The Making of Urban America* (Princeton University Press, 1965) and *Cities of the American West* (Princeton University Press, 1979), survey the origins of Chicago, Denver, and numerous cities in between. Regional architectural history is developed in Carl W. Condit, *Chicago, 1910–1929: Building, Planning and Urban Technology* (University of Chicago Press, 1973); and in Paul Clifford Larson and Susan M. Brown, *The Spirit of H. H. Richardson on the Midland Prairies* (Iowa State University Press, 1988). Regional landscape design and conservation history is discussed in Robert E. Grese, *Jens Jensen: Maker of Natural Parks and Gardens* (Johns

Hopkins University Press, 1992), which is part of the award-winning series of books, Creating the North American Landscape.

Collections of regional/historical studies of the Great Plains include Brian W. Blouet and Merlin P. Lawson, editors, *Images of the Plains: The Role of Human Nature in Settlement* (University of Nebraska Press, 1975); Brian W. Blouet and Frederick C. Luebke, editors, *The Great Plains: Environment and Culture* (University of Nebraska Press, 1980). Specific regional treatments are too numerous to list, but include: Timothy R. Mahoney, *River Towns in the Great West: The Structure of Provincial Urbanization in the American Midwest, 1820–1870* (Cambridge University Press, 1990); Richard G. Bremer, *Agricultural Change in an Urban Age* (University of Nebraska Studies, No. 51, 1976), a study of agricultural adjustment in the Loup Valley of Nebraska; and C. Barron McIntosh, "Patterns from Land Alienation Maps," *Annals,* 66 (1976), pages 570–582, which outlines the course of land settlement in the Nebraska Sand Hills. If one is interested in the rise and fall of the range cattle industry from 1845 to the turn of the twentieth century, see Ernest Staples Osgood's classic account, *The Day of the Cattleman* (University of Minnesota Press, 1929).

American travel guidebooks have yet to surpass the volume produced by the Works Progress Administration/Federal Writers' Project in the late 1930s; some have been reissued with new introductions, including: Neil Harris and Michael Conzen, *The WPA Guide to Illinois* (Pantheon, 1983). Field trips around Denver and the Colorado Piedmont are outlined in Steven L. Scott and Charles O. Collins, editors, *Field Trip Guide, 1983 Denver Meeting* (Association of American Geographers, 1983). For Iowa see D. Ray Wilson, *Iowa Historical Tour Guide* (Crossroads Communications, 1986). A wealth of information on things to see and do in Illinois, Iowa, Nebraska, and Colorado is provided in *Traveling Interstate 80 with Otto* (Travel Guide Publications, 1990).

A number of fine book series have been published over the last fifty years, in addition to the WPA/FWP guides mentioned above. These include *Rivers of America,* a series of about fifty books published during the 1940s and 1950s by Farrar & Rinehart and Rinehart; *The States and the Nation,* a series of fifty books published for the national Bicentennial of the American Revolution by W. W. Norton and the American Association for State and Local History; and *The Smithsonian Guide to Historic*

America, a series of twelve volumes organized by regions and published in 1989–1990 by Stewart, Tabori, and Chang. Within *Rivers of America, The Missouri* by Stanley Vestal (Farrar & Rinehart, 1945) is especially pertinent; but also of use are *Upper Mississippi* by Walter Havighurst, *The Chicago* by Harry Hensen, and *The Illinois* by James Gray. Within *The States and the Nation* see *Michigan* by Bruce Catton (1976); *Indiana; A History* by Howard H. Peckham (1978); *Illinois: A History* by Richard J. Jensen (1978); *Iowa* by Joseph F. Wall (1978); *Nebraska* by Dorothy Weyer Creigh (1977); and *Colorado: A History* by Marshall Sprague (1976). Within *The Smithsonian Guide to Historic America* see Volume VI, *The Great Lake States: Ohio, Indiana, Illinois, Michigan, Wisconsin, Minnesota* by Suzanne Winckler (1989); Volume VII, *The Plains States: Missouri, Kansas, Nebraska, Iowa, South Dakota, North Dakota,* by Suzanne Winckler (1990); and Volume VIII, *Rocky Mountain States: Colorado, Wyoming, Idaho, Montana,* by Jerry Camarillo Dunn, Jr. (1989).

Two of the greatest works of twentieth-century American fiction are located in Nebraska: *My Ántonia* by Willa Cather (Houghton Mifflin, 1949); and *The Works of Love* by Wright Morris (Alfed A. Knopf, 1952). Two great historical works are *The Oregon Trail* by Francis Parkman (Putnam, 1849); and *Across the Wide Missouri* by Bernard DeVoto (Houghton Mifflin, 1947). DeVoto's edition, *The Journals of Lewis and Clark* (Houghton Mifflin, 1953), is also noteworthy.

And two works of photography deserve mention. They are Robert Adams, *From the Missouri West* (Aperture, 1980); and Frank Gohlke, *Measure of Emptiness: Grain Elevators in the American Landscape* (Johns Hopkins University Press, 1992), which is part of the Creating the North American Landscape series.

Index